India Bangladesh Domestic Politics

Punam Pandey

India Bangladesh Domestic Politics

The River Ganges Water Issues

Springer

Punam Pandey
Institute for Reconciliation and Social Justice
University of the Free State
Bloemfontein
South Africa

ISBN 978-981-10-2370-5 ISBN 978-981-10-2371-2 (eBook)
DOI 10.1007/978-981-10-2371-2

Library of Congress Control Number: 2016947501

Printed on acid-free paper

This Springer imprint is published by Springer Nature
The registered company is Springer Science+Business Media Singapore Pte Ltd.

To my parents

Preface

Conjoined by history, economy, culture, geography and environmental concerns, the relationship between India and Bangladesh has been defined mostly by domestic politics. Politics on the Ganges seems to have contributed to a political discourse in Bangladesh where the river became a yardstick to judge the performance of a government for quarter of a century. This hydrological issue began when Bangladesh was part of Pakistan and continued even after its birth as an independent country in 1971.

After two decades of deliberations with Pakistan and a quarter of a century of negotiations with Bangladesh, the treaty on the Ganges was signed in 1996. Though lauded for a successful conclusion of one of the contentious river disputes all over the world; it is important to analyse the variables which kept the issue alive for such a long time, and address new challenges which would be determining the life of the treaty after 2026 when the term of the treaty ends. Examining the operation of the treaty is equally important. These questions spurred me to go for detailed examination of the subject.

There are some iconic works on the subject like Ben Crow's *Sharing the Ganges: The Politics and Technology of River Development*; B.G. Verghese's *Waters of Hope: Integrated Water Resource Development and Regional Cooperation within the Himalayan-Ganges-Brahmaputra-Barak Basin,* and Ramaswamy Iyer's *Water: Perspectives, Issues, Concerns*. These are detailed books on the Ganges, giving a historical treatment to the subject; describing evolution of the issue in bilateral relationship; and also suggesting how the river can contribute to regional cooperation. The present work studies the Ganges from domestic political lens and understands how this river occupied the focus of discussion for major part of the bilateral relationship of India and Bangladesh. This book has emerged out of a prolonged research on the issue.

For deeper analysis of the subject, I have gone through coverage of the issue in mass media of India and Bangladesh for almost four decades.[1] To understand the

[1]Though Calcutta has become Kolkata, the old names have been used keeping the context in focus.

Indian political discourse on the Ganges, I have also consulted Parliament's proceedings. Reading through the Indian parliamentary discussions, I found that they underline that conventional understanding has developed over the period among the political parties that water issue with neighbours should be dealt bilaterally. Another noticeable feature is the gradual rise of sharp disjuncture in federal relationship where states cannot be assumed to be on board when the central government is negotiating with neighbour on any issue especially related to water. The central argument of the book is *how domestic politics facilitate or complicate external negotiation process in the context of the Ganges river issue*. The book approaches the Ganges river water sharing issue with fresh research insights learnt by fieldwork both in India and Bangladesh.

Though the Ganges region is home to one of the densest concentrations of some of the poorest people in South Asia, flexibility and sensibility required to resolve this issue was not sufficiently demonstrated by both countries. Achieving immediate political interests has always preceded over long-term perspectives. The reading of important events highlight that there were occasions when matter would have been resolved for the larger good of the people but political considerations overwhelm the governments of the day. Both sides stuck to their respective positions. Bargaining and compromise were made possible only when linkage among various issues were aligned; sometimes it was just for fulfilment of domestic political configurations. The 1977 Agreement and 1996 Treaty underline that the only missing link between negotiation and successful outcome on previous occasions has been lack of convergence of political interests. It highlights that political dimension works as a master key which could open all the hitherto closed doors. This has been illustrated in the book with several examples. Another important aspect which emerges is that technical differences are circumvented easily if overlaid by interpretations according to alignment of interests of the political parties in power which, in turn, work as a multiplying factor for them in the domestic politics.

Though the Ganges Treaty has covered close to two decades of its 30 years period, importance of the subject has not diminished for these two South Asian neighbours; rather it has increased because of climate change and other emerging factors. Reading of the Ganges dispute suggests that river water in general and Ganges in particular has political currency in the domestic politics of both India and Bangladesh. This aspect has been gleaned from an examination of various phases of negotiations. The discussions on domestic aspects are critical because they are undercurrents which guide the respective governments and its negotiating team. Choice of negotiating strategy has always been contentious between India and Bangladesh; while the former prefers bilateralism and the latter intends to follow multilateralism. On its trajectory to formalisation of the treaty, civil societies played very important roles. Their important contributions in the Ganges water negotiation have not been sufficiently highlighted in the literature. This book examines the constraints and limitations faced by civil society groups which are engaged between two neighbours and discusses the future potential of civil society in the river water dispute. It is important to examine operation of the Ganges Treaty for last two decades which would help facing new challenges of climate change and other evolving political situation of India and Bangladesh.

The study of this subject becomes critical because this is the younger region in terms of modern nation-state demands where nation-building projects are still in the fledging stage, basic needs of people are yet to be realised; many more negotiations would be carried out as India and Bangladesh share another 53 rivers where same considerations would play many times over. Most importantly, climate change phenomenon has come out of the realm of possibility to reality as different parts of world in general and India in particular are suffering from successive droughts. But for many Indian provinces, it seems that confrontation and combativeness have taken over debates and deliberations over issues of common concerns particularly for those that share water resources. Recently, there have been some fresh initiatives to cleanse the Ganges and activate pause button on the construction of large dams on the upper reaches of the river. The result would be known in future whether all these measures add to the free flow of the river, and hopefully, this would translate into better riparian relationship between India and Bangladesh.

In the end, it is important to add one caveat for researchers who are interested in studying Indo-Bangladesh river water relationship that approaching government authorities for access to data and getting updates about status of bilateral relationship is next to impossible. One has to be extra-vigilant and use informal sources to get important information. This book will be of interest to students studying international conflict over rivers, student and scholars of South Asia and international relations, scholars and practitioners of Indian foreign policy, journalists, water experts and civil society groups.

Bloemfontein, South Africa Punam Pandey

Acknowledgement

Accomplishing any project is made possible because of assistance, cooperation and encouragement of well-wishers. This is the place to express my gratitude. First and foremost, I would like to appreciate Prof. Andre Keet of the Institute for Reconciliation and Social Justice, University of the Free State, South Africa, for tremendous support that has been invaluable in finalising this work. I owe special thanks to Prof. Keet for giving me the liberty to pursue this project.

Also deserving my gratitude is Prof. Navnita Chadha Behera (Professor, Department of Political Science, University of Delhi), who always encouraged and motivated me. Her regular nudging keeps reminding me that I have to complete this project. Two important souls who were always generous to me with their time and scholarship are no more, but this work would be incomplete without acknowledging their names, late B.G. Verghese and Ramaswamy Iyer.

Undoubtedly, this work would not have been possible without support of some important people from both India and Bangladesh. I would also like to thank Professor Achin Vanaik, Ambassador Rajendra Abhyankar, Dr. Sreeradha Datta, Prof. Veena Sikri (former High Commissioner of India to Bangladesh) and Dr. Smruti Pattanaik. I got tremendous support from Prof. Imtiaz Ahmed, Prof. Dalem Burman, Prof. Ameena Mohsin, Ambassador Ashfaqur Rahman, and Prof. Asif Nazrul. For the project, I interviewed policymakers, former ministers, members of Joint River Commission, academicians, members of civil society and think tanks in India and Bangladesh between 2009 and 2011 on regular intervals. I also consulted the government functionaries who have been actively associated with the Ganges water negotiation.

Apart from this, as part of track-II group, I was privileged to meet senior and retired bureaucrats, former ministers, activists, academicians, media personnel and members of advocacy groups closely. Being a part of these groups has been a great learning experience. Through these discussions, conversations and field experiences, I have been able to sharpen my arguments and refine ideas, some of which have been presented in earlier works. Sections of this book draw on material from my following publications: "Bangladesh, India and Fifteen years of Peace: Future

Directions of the Ganges Treaty" (*Asian Survey*) and "Revisiting Politics of the Ganga Water Dispute between India and Bangladesh" (*India Quarterly: A Journal of International Affairs*). I also received useful feedback on an earlier version of third chapter, "Negotiation on the Ganges: Idealism of Regional Cooperation or Pragmatic Bilaterialism" presented at National University of Singapore.

Lastly and most importantly, staff members of many libraries have especially been really kind, helpful and patient in locating papers, books, newspaper archives. Some of these libraries are Parliament Library, The Nehru Memorial and Museum Library (at Teen Murti House), and libraries of Institute for Defence Studies and Analysis (IDSA), Jawaharlal Nehru University (JNU) and Indian Council of World Affairs (ICWA). For any insufficiency of facts and interpretations, sole responsibility lies with me.

Contents

1 Cooperation to Confrontation to Cooperation over the Ganges Between India and Bangladesh 1
1.1 Historical Background: The Legacy of the Pakistan Era 2
1.1.1 Mutual Distrust and Divergent Negotiating Strategies 4
1.1.2 Domestic Politics: A Victim of Neglect 5
1.2 Indo-Bangladesh Negotiations on the Farakka Barrage 6
1.2.1 First Phase: Friendly Regimes and Productive Talks 7
1.2.2 The Domestic Pressures 8
1.2.3 Second Phase: A Total Rupture 9
1.2.4 Third Phase: Rapprochement and the First Agreement on Water Sharing 11
1.2.5 Fourth Phase: The See-Saw Game: Reverting to the Old Paradigm 13
1.2.6 The Domestic Pressures 15
1.2.7 Bilateral Talks: Turning a New Leaf? 16
1.2.8 Farakka: The Political 'Football' in Bangladesh's Domestic Politics 18
1.2.9 Fifth Phase: The Final Breakthrough 22
1.2.10 A Critical Evaluation of the Functioning of the Treaty 25
1.3 Conclusion 30
Annexure 31
References 35

2 Domestic Variables: Compulsions of Competitive Politics 37
2.1 Shared History and Divergent Aspirations 38
2.2 Compulsions of Indian Federalism 41
2.2.1 West Bengal Being Conscious of Its Position 41
2.3 India as a Factor in Bangladesh Politics 47
2.4 Mixing up Internal and External Relationship 50

2.4.1 India Synonymous of Mujib Government ... 50
2.5 Assertiveness of Muslim Identity ... 54
2.5.1 Complete Reversal ... 54
2.5.2 Farakka Discussion: Manoeuvring of Politics ... 57
2.6 Conclusion ... 62
Annexure ... 63
References ... 67

3 India's Negotiating Strategy on the Ganges: Bilateralism Versus Multilateralism ... 69
3.1 Evolution of India's Negotiating Strategy on River Waters ... 70
3.1.1 India and Bangladesh Negotiations on Farakka Barrage ... 71
3.2 Bangladesh's Attempt to Internationalise the Ganges River Water Issue ... 74
3.3 Interregnum of Multilateralism in Bilateral Negotiation ... 81
3.3.1 Acceptance of Nepal Option by India ... 81
3.4 Bilateralism Versus Multilateralism ... 83
3.5 Conclusion ... 88
Annexure ... 89
References ... 97

4 Role of Civil Society in the Ganges Water Negotiations ... 99
4.1 Introduction ... 99
4.2 Defining Civil Society ... 100
4.2.1 Role of Civil Society in Resolution of the Ganges River Water Dispute ... 100
4.2.2 Constraints and Limitations of Role of Civil Society ... 103
4.3 Future Potential of Role of Civil Society ... 107
4.4 Conclusion ... 107
References ... 108

5 The Ganges Treaty: Future Directions ... 109
5.1 Experience of 15 Years of the Ganges Water Treaty ... 109
5.2 Operations of Joint Rivers Commission (JRC) ... 113
5.3 Climate Change and Impact on Ganges River Water Treaty ... 119
5.4 Interlinking of River Projects and Bangladesh's Concerns ... 122
5.5 Emerging Role of Provinces in Indian Federation ... 122
5.5.1 Exploration of Alternatives ... 125
5.6 Conclusion ... 127
References ... 128

6 Analytical Observations and Conclusion ... 131
References ... 135

Index ... 137

About the Author

Punam Pandey is currently Postdoctoral Research Fellow at the Institute for Reconciliation and Social Justice (IRSJ), University of the Free State, South Africa. Before that she worked as Assistant Professor at the Department of Political Science, University of Delhi, India, from where she also obtained her Ph.D. As a postdoc researcher at the IRSJ, she has written extensively on water politics from the perspective of international relations. She broadly works on non-traditional security issues in South Asia with a special focus on India and Bangladesh. Some of her publications can be accessed in the refereed journals like *Asian Survey*, India Quarterly: *A Journal of International Affairs*, and *South Asian Survey* among others. Her current research project at the IRSJ is on the issues of memory contestation where she researches the Liberation Movement in the competitive politics of Bangladesh.

Abbreviations

AL	Awami League
BAKSAL	Bangladesh Krishak Sramik Awami League
BIE	Bangladesh Institute of Enterprises
BISS	Bangladesh Institute of International and Strategic Studies
BML	Bangladesh Muslim League
BNP	Bangladesh Nationalist Party
BSS	Bangladesh Sangbad Sangstha
BUP	Bangladesh Unnayan Parishad
CPD	Centre for Policy Dialogue
CPI (M)	Communist Party of India (Marxist)
CPR	Centre for Policy Research
ECAFE	Economic Commission for Asia and the Far East
FAO	Food and Agricultural Organisation
GDP	Gross Domestic Product
IDSA	Institute for Defence Studies and Analysis
IISD	International Institute of Development Studies
JCE	Joint Committee of Experts
JRB	Jatiyo Rakkhi Bahini
JRC	Joint River Commission
MoU	Memorandum of Understanding
MP	Member of Parliament
NGO	Non-Government Organisation
ORF	Observer Research Foundation
SAARC	South Asian Association for Regional Cooperation
UBDC	Upper Bari Doab Canal
UN	United Nations
UPA	United Progressive Alliance
USAID	United States Agency for International Development
USSR	Union of Soviet Socialist Republics
WMO	World Meteorological Organisation

Chapter 1
Cooperation to Confrontation to Cooperation over the Ganges Between India and Bangladesh

The Ganges has determined the health of the bilateral relationship of India and Bangladesh for almost quarter of a century. Before 1971, Bangladesh constituted an eastern wing of Pakistan. The Farakka barrage, constructed to divert water from the Ganges to Hooghly River, was a bone of contention even between India and Pakistan. Thus, Bangladesh inherited this legacy from Pakistan, and accorded importance to sharing of the Ganges river water from its birth. Since then, the river stayed on priority list of bilateral issues between two neighbours till 1996. Furthermore, this subject became imbedded into domestic politics of Bangladesh to such an extent that Dhaka's handling of this subject also determined the fate of its top political leaders and, considered as the single most important yardstick for judging the performance of its ruling regimes.

This chapter describes different phases of the bilateral negotiations on the Ganges between India and Bangladesh. The chapter further argues that both countries' agreements, or indeed disagreements have *not* been shaped so much by the technical issues at stake but other considerations. In fact, predisposition of reaching an agreement on this matter is directly linked to two critical factors: first, the political relationship between their respective ruling regimes in New Delhi and Dhaka at any given historical juncture; and second, the politicisation of the Farakka issue, especially in Bangladesh, by various political parties and leaders for their domestic political ends.

The chapter is broadly divided into two parts. First entails a brief overview of the historical legacy of the Pakistani regimes' approach in the external and internal domain towards negotiations with India on the Farakka barrage; and the second part presents a much more detailed discussion on the post-1971 phase between India and Bangladesh. This part is further divided into five phases, which takes into account

---

Originally published in India Quarterly: A Journal of International Affairs, Vol. 68 No.3

P. Pandey, *India Bangladesh Domestic Politics*,
DOI 10.1007/978-981-10-2371-2_1

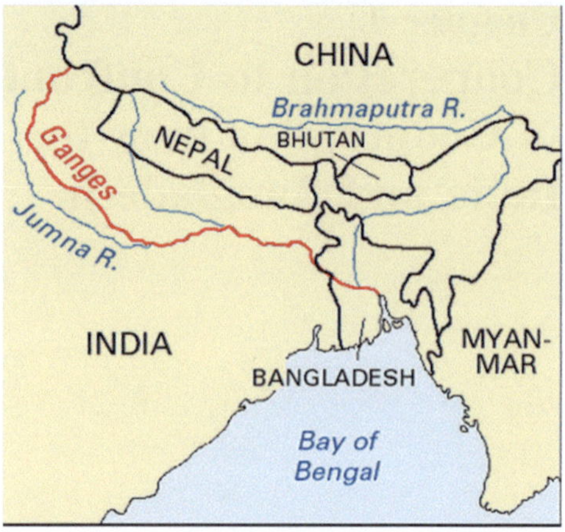

Fig. 1.1 Ganges in the region

the changing equations of New Delhi's relationship with Dhaka and how these have, in turn, influenced the negotiations on the river water sharing. Before discussing historical progression and political dynamics, it is important to briefly outline the geographical features of the Ganges river.

The Ganges rises in the foothills of the *Gangotri* on the southern slope of the Himalayan range in India and moves in a south-east direction towards Bangladesh. Seven major tributaries augment its flow on its way. The largest portion of the Ganges falls within India. Within India, this is an intra-state river shared by the provincial states of Uttar Pradesh, Himachal Pradesh, Rajasthan, Haryana, Madhya Pradesh, Bihar, West Bengal and the National Capital Territory of Delhi. The mainstream of the Ganges bifurcates into two channels which are known as Bhagirathi-Hooghly in India and Padma in Bangladesh. Calcutta city, the capital of West Bengal and one of India's most important ports, is situated on the Hooghly River. In South of Calcutta, the Hooghly is joined by the Damodar (tributary of the Ganges) river. The Ganges traverses a 2510 km long journey through India and Bangladesh. This hydrological feature divides India and Bangladesh as upstream and downstream riparian countries respectively. The river itself flows for over 92 % of its course within India before joining the Brahmaputra and Meghna rivers in Bangladesh, the combined river flows into the Bay of Bengal. Out of the total drainage area of the Ganges 79 % belongs to India, 4.2 % to Bangladesh and almost 14 % to Nepal (Fig. 1.1).

1.1 Historical Background: The Legacy of the Pakistan Era

The genesis of the Farakka barrage goes back to the British era who were beginning to face navigational difficulties to the port of Calcutta. The Bhagirathi used to carry the bulk of the Ganges flows but by the late eighteenth century, the Padma had become the main carrier channel of its waters. The deterioration of Bhagirathi's water levels seemed to be a threat not only to the shipping of the Calcutta port but also challenged its very survival.[1] The British commissioned several studies including those carried out by Sir Arthur Cotton (1853), Vernon-Harcourt (1896), Reak (1913), the Stevenson-Moore Committee (1916–1919), Sir William Wilcocks (1930), T. M. Oag (1939), A. Webster (1946) which recommended the construction of a barrage across the Ganges and regulation of its headwater supplies in order to save the Calcutta port since it was one of the most important ports for India as well as for the overseas trade of landlocked Bhutan and Nepal.[2] After independence, the Indian government commissioned a study to resolve the problem of low water level in Bhagirathi-Hooghly, which also recommended construction of a barrage at Farakka. The diminishing headwater supply of Bhagirathi-Hooghly especially in the dry season period was silting the water ways and rendering it inaccessible for big ships. When Pakistan learnt about India's proposal to construct the Farakka barrage, it protested and sought immediate consultations as the proposed project (Pakistan believed) would have far-reaching consequences for the agricultural economy and commercial life of East Pakistan. This germinated another bilateral tension between India and Pakistan, already fragile on account of the Kashmir conflict and the division of Indus water resources. Both countries held a bilateral dialogue to resolve their differences over this issue but in vain. The following sections describe the negotiating process between the two countries and also analyse the causes of its failure.

When India was contemplating different proposals to address the navigability issue of the Calcutta port, Walter Hensen commission recommended the construction of the Farakka project as the "only purposeful measure" for the preservation of Calcutta port in 1957.[3] Meanwhile from 1952 to 1955, Pakistan moved different project proposals such as the Ganges-Kobadak project. India responded with alternative propositions including that of the Farakka barrage. In 1956, India and Pakistan agreed to cooperate on the river water projects on the Ganges[4] and these negotiations got underway in 1960. Four rounds of technical experts meeting were held from June 1960 to January 1962, alternating between India and Pakistan. Since India agreed to address Pakistani concerns about the Farakka project and inquired about its potential water requirements from the Ganges, the latter offered widely varying accounts ranging from 2,000 cusecs in the first meeting to 49,000 cusecs in the last meeting; it was ever increasing. In 1954 meeting, for example, the quantum of water demanded was 2000 cusecs, which rose gradually to 49,000 cusecs in 1968 in just 14 years. Pakistan had perhaps not clearly formulated any plans but since India sought justification for the latter's water requirements, the former began to draw up very different plans for dams.[5] India on its part "unwisely questioned every such demand for water

[1]Rangachari and Iyer (1994), pp. 173–174.

[2]Ibid.

[3]Ibid., p. 174.

[4]Ibid., p. 175.

[5]Personal interview with the author at New Delhi on 27 August 2009.

even when this was initially pitched at no more than 3500 cusecs and insisted on seeking more data and clarifications. By 1968, Pakistan's demand was inflated to such an extent that it demanded more than the entire 55,000 cusecs low season flow of the Ganges—a proposition Bangladesh later inherited and made its own, convinced that Farakka diversions were irrelevant to the Calcutta port and were basically "an evil design to harm it".[6] The reason why bilateral negotiations between India and Pakistan failed to resolve this issue is explained in the following section.

1.1.1 Mutual Distrust and Divergent Negotiating Strategies

India and Pakistan were born as hostile neighbours in 1947. This was partly due to the partition and its horrific aftermath in terms of huge losses of lives; the ensuing war on the question of Kashmir; and partly due to their incongruent self-vision and status in the regional power matrix. Pakistan believed that British India was divided into two parts and hence, always sought juridical equality and power parity with India, while the latter wanted to assert its *numero uno* position in the natural hierarchy of South Asia.[7] This seemed to shape their respective understanding and negotiating strategies.

Gradually, both countries decided to have discussions of water resource experts on projects of mutual interest. Even when water expert-level discussions were on, in 1961, Pakistan's president Ayub Khan wrote to the Indian Prime Minister Jawaharlal Nehru for elevating talks at the ministerial level. Indian Prime Minister, however, stressed the need to have a practical agenda before undertaking a ministerial talk and underlined the need for a full exchange of data. However, both countries agreed that this was a human problem and should be dealt with by taking into account both parties' needs and difficulties.[8]

For 5 years, there was a lull between the two countries regarding water negotiations because of the Indo-China war in 1962 and the Indo-Pakistan war in 1965. The 1965 war had ended with the Tashkent Agreement. The Agreement was concluded between Lal Bahadur Shastri, Indian Prime Minister, and President Ayub Khan of Pakistan on January 10, 1966 in the presence of the Soviet Premier, Alexei Kosygin. Pakistan approached super powers and sought their mediation for resolution of the bilateral issue. Initially, the former Soviet Union responded positively and its Premier Alexei Kosygin visited Pakistan in April 1968, giving a boost to their bilateral ties. But soon India and Soviet Union came closer.

---

[6]Verghese (1999), pp. 391–392.

[7]Muni (1980).

[8]Rangachari and Iyer (1994), p. 176.

Meanwhile, bilateral talks resumed in 1967 and five secretaries' level meetings were held between December 1968 and July 1970. Pakistan, once again, resorted to making wide-ranging projections of water requirements while recording its reservations about the Indian projects. At the end of its fifth meeting, India concluded that Pakistan's water needs from the Ganges were in reality far modest than its claims and it did not adversely impact its interests while India was overwhelmingly dependent on it for its agricultural and other requirements. This was in consonance with Pandit Nehru's assessment, which he had shared with the Parliament much earlier in a statement that: "it is our view that there should be no real injury caused to Pakistan by the scheme".[9] India offered to quantify water which could be released to Pakistan if the latter proposed more realistic projects. Pakistan disagreed and argued that it must have an idea of the "guaranteed supplies" of water before any reformulation of the Farakka Barrage project. Pakistan pressed for ministerial level talks again in 1970 but India insisted that secretary level meetings needed to continue till appropriate preparations were completed for a higher level meeting.[10] In the last of the Secretariat meetings, in July 1970, some progress was made. The two sides agreed that "the point of delivery of supplies to Pakistan of such quantum of water as may be agreed upon will be Farakka".[11] India refused ministerial meeting because general election was due in India in 1971, which were won by the Congress party led by Mrs. Indira Gandhi who refused to concede the demand for a ministerial level meeting and instead proposed another round of secretary level talks. Though Pakistan agreed but internal situation in East Pakistan deteriorated very fast and this crisis led to a division of Pakistan with the liberation of its eastern wing as an independent state of Bangladesh.

1.1.2 Domestic Politics: A Victim of Neglect

It is well established that eastern Pakistan was marginalised politically, socially and economically within the Pakistani polity ever since its creation in 1947. East Pakistan experienced internal colonisation at the hands of the 'Punjabi' ruling regimes ensconced in the western wing.[12] As a clear derivative of the situation, the Farakka barrage was marginalised. Though India's announcement of constructing this barrage had evoked Pakistan's protests, as compared to other conflicts over

[9]See the reply by the Prime Minister, Shri Jawaharlal Nehru on Farakka Barrage, Second Series, vol. LVI, 19 August 1961, 7 to 19 August 1961, p. 3210.

[10]Rangachari and Iyer (1994), p. 177.

[11]Crow et al. (1995), p. 93.

[12]For detailed discussion on this issue, see Hasan (1994), Ahmed (2001), Islam (2003), Khan (1975).

river water sharing such as the Indus river, the Farakka barrage was always pushed to a secondary position by Pakistan, the rational being that the character of these two conflicts was somewhat different.

West Pakistan had a large irrigation system which depended on the Indus waters for its supplies. These were cut off by India just after partition. As a result, there was an imminent danger of war between the two countries. In East Pakistan, no such threat of war existed at that time and there was no organised irrigation system to contend about. The Farakka barrage was still distant. However, the timing of the Indus Basin Treaty signing between President Ayub and Pandit Nehru could have been propitious for an understanding between the two countries regarding the Ganges water also *but that opportunity was not availed of.*[13] The Farakka was relatively of much lesser importance. During the Indo-Pak war of 1965 too, President Ayub was preoccupied with the question of Kashmir and the West Pakistan. He had, in fact, left East Pakistan completely undefended and without any military cover on the grounds that the defence of the eastern wing lay in the offensives mounted from the West, which shocked the Bengalis. At the same time, the Awami League leaders recognised that India made no military incursion upon their province, taking it as friendly discrimination in favour of East Bengal.[14] The internal politics of Pakistan was directed to use the Farakka barrage issue as matter of secondary importance. Kashmir issue has been of priority for West Pakistan. B. M. Abbas has described this with an illustration. During the Zulfikar Ali Bhutto and Swaran Singh meeting held in Islamabad on March 1, 1966 after the Tashkent declaration, "the meeting was deadlocked at the very start on account of a procedural wrangle over the proposed agenda. The Indian side wanted a step by step approach to deal with the various problems but Pakistan insisted on discussing the Kashmir question first. So the Farakka or any other problem could not be considered at all".[15]

1.2 Indo-Bangladesh Negotiations on the Farakka Barrage

The creation of Bangladesh marked a significant point of departure in the resolution of the Farakka barrage dispute. The following section describes all phases of Indo-Bangladesh negotiation on the Farakka.

[13]Abbas (1984), pp. 27–28.

[14]Chakravarthtty (1986), pp. 274–278.

[15]Abbas (1984), p. 27.

1.2.1 First Phase: Friendly Regimes and Productive Talks

Bangladesh's first Prime Minister, Sheikh Mujib ur Rahman appreciated and duly acknowledged India's role in liberating Bangladesh and sought to establish very friendly relations with it.[16] In March 1972, a Treaty of Friendship, Cooperation and Peace was signed which envisaged among other aspects joint studies and an action plan for flood control, river basin development and the development of hydropower and irrigation.

The first round of political dialogue was held in Delhi between India's Minister of External Affairs, Sardar Swaran Singh and his Bangladesh's counterpart, Khondakar Moshtaque Ahmed on July 8, 1973, reaffirming that a final decision on the sharing of the Ganges would be taken at a summit meeting between the two Prime Ministers scheduled to be held in 1974. B.M. Abbas, who had been participating in the water negotiations as a Member of Pakistan's team before 1971 recalls a conversation with India's Foreign Minister, Sardar Swaran Singh. While discussing India's decision to supply coal for setting up a power station in Bangladesh, "Mr. Swaran Singh saw no difficulty in increasing power potential of Karnafuli as lands to be submerged within India were barren and I pointed out that India had objected to this project in Pakistan days and alleged submersion of an Indian copper mine, he laughed and said that *the situation was now different*".[17] (italics added) This shows that if ties between the ruling regimes are friendly, it helps develop a mutual understanding between the neighbours. This was evident from joint declaration issued at the end of the visit of the Bangladesh's Prime Minister Sheikh Mujibur Rahman to India in May 1974 wherein the two sides had for the first time underlined the need for arriving at a mutually acceptable agreement and recognising the real obstacle in terms of the flow of Ganges not being sufficient to meet the needs of both sides. It is worthwhile to cite from the text of the Joint Declaration in detail:

> 17. the two Prime Ministers took note of the fact that the Farakka Barrage Project *would be commissioned before the end of 1974*. They recognized that during the periods of minimum flow in the Ganges, there *might not be enough water to meet the needs of the Calcutta port and the full requirements of Bangladesh and therefore, the fair weather flow of the Ganges in the lean months would have to be augmented to meet the requirements of the two countries. It was agreed that the problems should be approached with understanding so that the interests of the both the countries are reconciled and the difficulties removed in a spirit of friendship and cooperation*. It *was accordingly decided* that the best means of such augmentation through optimum utilisation of water resources of the region available to the two countries should be studied by the Joint Rivers Commission. The Commission should make suitable recommendations to meet the requirements of both the countries...

---

[16]For India's role in Bangladesh liberation see Ayoob and Krishnaswamy (1972) Wilcox Wayne (1973).

[17]Abbas (1984), pp. 30–1.

> 18. it was recognised that it would take some years to give effect to the recommendations of the Commission as accepted by the two Governments. In the meantime, the two sides expressed their determination that before the Farakka Project is commissioned *they would arrive at a mutually acceptable allocation of the water available during the periods of minimum flow in the Ganges.*[18] (italics added)

The Joint Rivers Commission (detailed functioning of the JRC will be discussed in the fifth chapter) exchanged the data and worked out availability of water at Farakka in different periods. Despite intense negotiations from June 1974 onwards, the JRC could not reach an agreement on augmentation of the lean season flow of the Ganges. While New Delhi suggested augmentation at lower Ganges from Brahmaputra in view of the limited storage potential on the Ganges and rapidly growing demands for agricultural requirements within India. Bangladesh, on the other hand, continued to insist on augmentation within the Ganges basin. Two ministerial level meetings took place in February and April 1975 but failed to break the stalemate on augmentation.

In April 1975 meeting, both India and Bangladesh reached an agreement that allowed experimental operation of the Farakka barrage.[19] Joint teams were to observe the effects of the diversions at Farakka, both on the Hooghly in India and inside Bangladesh.[20] The two sides further agreed that India would carry varying discharges ranging from 11,000 to 16,000 cusecs in 10-day periods from April 21 to May 31.[21] But Bangladesh felt betrayed by the way the interim agreement worked out to withdraw water at Farakka; the barrage was commissioned pending any headway on a mutually acceptable solution. This strain was reflected when Bangladesh's Water Resources Minister Abdur Rab Serniabat cancelled the visit to attend the commissioning ceremony.[22]

1.2.2 The Domestic Pressures

The opposition in Bangladesh and Mujib's critics played a major role in undermining the goodwill between the two countries. His adoption of policies of secularism and socialism were understood as a consequence of influence of India and

---

[18]Extract of para 17 and 18 from the Joint Declaration of the Prime Ministers of India and Bangladesh in New Delhi on 16th May, 1974, as cited in Avtar (1996).

[19]The Farakka barrage is designed to serve various needs of West Bengal. Most important reason for construction of the barrage is to preservation and maintenance of Calcutta port by improving the regime and navigability of the Bhagirathi-Hooghly river system. This also provides a communication link between North Bengal, North-East India, Sikkim and Bhutan with the rest of India through rail-cum-road bridge built across the river Ganges at Farakka. http://mowr.gov.in/writereaddata/linkimages/anu141200246187.pdf, accessed on 14 July 2012.

[20]Crow et al. (1995).

[21]Tariq (1998), p. 222.

[22]Iftekharuzzaman (1994), pp. 221–2.

Soviet Union. This was severely criticised in the Bangladesh polity and fuelled anti-Indian sentiments. Background details of opposition to Mujib regime will be discussed in the next chapter.

Bangladesh's National Awami Party leader Maulana Bhashani opposed the Friendship Treaty and criticised Mujibur Rahman for having mortgaged the political and economic sovereignty of Bangladesh to India. The unresolved issue of the Farakka further fuelled such sentiments. The former Ambassador, Harun ur Rashid explains that Sheikh Mujib was facing a tough time at the home front. People were dismayed with the introduction of one party rule and criticised Mujib for neither ensuring democratic governance nor was he able to get water from India.[23] Soon, there was an army coup spearheaded by a few disgruntled army officers. The army uprising assassinated Sheikh Mujib and his family members, overthrew the government and established a military regime headed by General Zia-ur Rahman.

1.2.3 Second Phase: A Total Rupture

The assassination of Mujibur Rahman quickly deteriorated the situation and simmering tensions came to the fore resulting in mutual distrust, and later an open conflict. In the years 1975 and 1976, the issue of the Farakka barrage was soon forgotten in the fallout of coups and countercoups in Bangladesh.

Though Bangladeshis strongly criticised the Indian government's decision to continue withdrawing water after 1975,[24] Indian side believes that "there was no 31 May deadline according to the agreement of 1975. It was mutually agreed that India would operate the Farakka barrage from April 21 to May 31. Anyway whatever agreement agreed between India and Bangladesh is, is valid only for lean period from January 1 to May 31. After May India is not subject to any obligation about quantity of water withdrawal".[25]

Mrs. Indira Gandhi had forged an excellent rapport and partnership with Mujib ur Rahman. In fact she viewed the liberation of Bangladesh as a victory for India's ideology and principles. This can be gauged with her statement after Bangladesh's independence "Our stand has proved correct".[26] But after Mujib's assassination, the bilateral relationship worsened quickly. According to an Indian government official who had been part of water negotiations, Joint River Commission was due to meet in August 1975; but the new military regime in Bangladesh had refused to participate in the meeting on the grounds that "India was supporting Kader Siddiqui

[23]Personal interview with the author at Dhaka on 27 October 2009.

[24]The accord was meant for the current lean season. But there is disagreement between India and Bangladesh to term it. While Bangladesh considers 31st May as deadline, India understand that this accord was for lean period.

[25]Personal interview of the author with an Indian government official who was associated with the Indo-Bangladesh water negotiations during that period at New Delhi on 23 May 2011.

[26]Mrs. Indira Gandhi as cited in Dhiren (1972), p. 94.

and other guerrillas who belonged to the Mukti Bahini at that historical juncture”.[27] In a total contrast, the dominant view in Bangladesh is that it was Mrs. Gandhi who was no longer interested in negotiation with new regime in Dhaka. Ambassador Harun concurs this point by citing a personal experience that when Bangladesh's new President, General Ziaur Rahman came to India and met Mrs. Gandhi; after meeting he described her as “a very difficult and tough lady”.[28] Later, Mrs. Gandhi's encounter with the Bangladesh's Water Resource Minister at New Delhi lasted for precisely five minutes because she was not keen to engage him on the Ganges Water issue. He simply believes that these incidents played an important role in alienating the Bangladesh's top leadership, which in turn pushed General Zia towards China and Pakistan which led deterioration in their bilateral relationship during Mrs. Gandhi regime and was also perceived as a move towards ganging up against India.[29]

At home, Zia-ur Rahman consolidated his position by winning the Presidential election held in 1978. Later, he formed the Bangladesh Nationalist Party (BNP) and held parliamentary elections in 1979. The BNP managed to mobilise the support of the Jamaat-e-Islami and the Muslim League constituency. After liberation, both these parties had been banned for collaborating with the Pakistani army. The emerging situation was remarkably different from that of the Mujib era. Ziaur Rahman amended the Constitution to replace the word Secularism with ‘absolute trust and faith in almighty Allah which should be “the basis of all actions”. Moreover, around the same time, Ziaur Rahman had allowed Ghulam Azam, leader of the Jamaat-e-Islam to come to Bangladesh in 1979 whose citizenship was cancelled by Mujibur Rahman for being complicit with Pakistan during the liberation struggle. His anti-India posture was also because after Mujib's assassination, some of the latter's supporters in the Jatiya Rakkhi Bahini had fled to India and operated from there to wage a guerrilla operation against his regime.[30]

On the water issue, Zia's government pursued a twin-track policy of holding India responsible for all its problems at home and seeking international support for its cause at various fora. The absence of any agreement on water sharing during 1976 and 1977s dry seasons provided fuel to the fire. A Bangladesh White Paper of 1976 asserted that the withdrawal of water by India threatened the survival of millions of people of Bangladesh; the media, with official encouragement, portrayed it as a “conspiracy against the independence and sovereignty of the country”[31] (More on this in third chapter). Bangladesh held Farakka responsible for all kinds of problems raging in the northern part of the country including the destruction of a bird sanctuary, floods and causing unemployment. But India reduced the flow into

---

[27]Personal interview of the author with an Indian government official who was associated with the Indo-Bangladesh water negotiations during that period at New Delhi on 23 May 2011.

[28]Personal Interview with the author at Dhaka on 25 October 2009. Also see Harun (2002), p. 62.

[29]Ibid.

[30]Ibid., p. 404.

[31]Khosla (2005), p. 73.

the Feeder canal below the minimum quantity required for flushing the Hooghly on March 27, 1976 as a gesture of goodwill and friendship. Meanwhile, the flow into Bangladesh increased further by a substantial amount during the current lean season. This fact is known to the high level technical delegation from Bangladesh which had visited Farakka on May 9, 1976.[32]

1.2.4 Third Phase: Rapprochement and the First Agreement on Water Sharing

The general election in India had for the first time brought a non-Congress government to power in 1977. Morarji Desai became the first non-Congress Prime Minister of India in three decades of independence and was keen to chart a different foreign policy from that of the Congress. A new bonhomie was established between the two regimes of India and Bangladesh in their very first meeting.[33] This broke the ice and not only paved the way for a new friendly relationship between New Delhi and Dhaka but also established a legacy that has survived till date. Thus, a non-Congress regime in New Delhi succeeded in establishing a good rapport with the BNP government in Dhaka, with far-reaching consequences for their bilateral relationship.

Indian Prime Minister Morarji Desai sent a high powered delegation to Dhaka led by the Minister of Defence, Mr. Jagjivan Ram, to hold talks over the Farakka dispute in April 1977. Both sides had struck to their respective rigid positions and a failure of these talks was only averted by Jagjivan Ram's decision to overrule his advisors. According to government official, talks were about to break down but Ram decided to overrule his advisers and make a generous allocation of 34,500 cusecs to Bangladesh.[34]

The Agreement had two parts: water sharing at Farakka and the long-term augmentation of the flows. The first related to the five lean months period from January 1 to May 31 in accordance with the agreed 10-day schedules, whereby if water flows were to fall down to 80 % of the assumed value in any 10-day period, India guaranteed Bangladesh 80 % of its stipulated share (For details see Annexure). A joint committee was constituted for the implementation of these arrangements. Two

[32]Statement of the Official Spokesman of the Ministry of External Affairs, Government of India on anti-Indian propaganda May 13, 1976 in New Delhi. Cited in "Sharing of River Waters", in Bhasin (1996), p. 402.

[33]When the Bangladesh Foreign Minister Shamsul Huq met the Indian Prime Minister, Mr. Desai for the first time, he did so without an appointment. When Desai said that "there should be no formality between neighbours", Huq remarked, 'Your Excellency, why have you stopped our rightful share of the waters of the Ganges?" to which Desai replied, "I did not do it. That Woman did". Cited in Khosla (2005), p. 74.

[34]Personal interview with a government official who was present during these negotiations in 1977. Also see, Maniruzzaman (1980), p. 209.

reviews were built in the Agreement at the end of 3 years and the other 6 months before the expiry of the treaty. The JRC was made responsible for investigating and studying schemes relating to economical and feasible long-term ways for augmentation of the dry season flows within a time frame of 3 years.

In 1978, India and Bangladesh exchanged their respective proposals for augmentation. India proposed the storages on the Dihang, Subansiri and Barak in the Brahmaputra–Meghna system, and a barrage at Jogighopa with a Brahmaputra–Ganges gravity link canal passing through both India and Bangladesh and joining the Ganges at a point just above Farakka. The Indian proposal considered the Ganges–Brahmaputra–Meghna Basin as an integrated system in which the densely populated Ganges sub-basin had relatively less water availability, less storage potential and greater demands for irrigation, whereas the sparsely populated Brahmaputra and Meghna sub-basins had relatively plentiful water availabilities, large storage possibilities and less demand for irrigation.

Bangladesh had reservations about the Indian proposal and argued that within India and Nepal the Ganges system had sufficient potential for augmenting flows at Farakka. It envisaged a number of storages in the upper catchments of the Ganges system for storing surplus monsoon flows and ensuring regulated releases so as to increase flows below Farakka. Bangladesh completely ruled out the possibility of any storage within its territory and also proposed to involve Nepal for purpose of data collection in this exercise. India and Bangladesh should, it was argued, seek Nepal's cooperation jointly as was decided in the exchange of side letters during the signing of 1977 Agreement (please see Annexure). Bangladesh believed that the massive transfer of water from the Brahmaputra to the Ganges basin as envisaged by the Indian proposal would lead to shortages in the Brahmaputra basin itself and had reservations about the link canal also.

The Indian side disagreed because they believed that these Bangladeshi propositions had not been investigated extensively. On technical grounds, both sides' position was not much different from the earlier period when Mujib and Mrs. Gandhi were negotiating the water-sharing arrangement at Farakka. That is perhaps because the technical experts and bureaucrats involved in their respective teams were largely the same. Mr. Reaz Rahman the former Bangladesh Foreign Secretary shared this with the author that the Congress party was especially opposed to the guarantee clause, which was an essential part of the 1977 Agreement.[35] He further added that in the 1980 election, the Congress Manifesto made the water treaty a part of its election plank and stated that India could not give away water to Bangladesh in the name of good neighbourly relations, and the treaty must be scrapped and millions of Bangladeshis who have crossed the border into Assam in order to flee its military regime must return. Congress promised to build a fence around its border with Bangladesh. But I went through the election manifesto of the Indian National Congress of 1980, and did not find any mention of the Ganges river water

[35]Personal interview with the author at Dhaka on 26 October 2009 at Dhaka.

Agreement of 1977. This reflects deep suspicion and prejudices present among government functionaries about each other.

1.2.5 Fourth Phase: The See-Saw Game: Reverting to the Old Paradigm

With the return of the Congress to power in New Delhi in 1980, the cordial atmosphere prevailing between two neighbours during Morarji Desai government dissipated, Ambassador Harun notes that in fact, when Bangladesh's Deputy Prime Minister came to New Delhi to personally deliver President Zia ur Rahman's letter to Prime Minister Mrs. Indira Gandhi, he waited for 3 days as a state guest at Ashoka Hotel in New Delhi because soon after his arrival, Mrs. Gandhi had left for an inter-state tour to *avoid* meeting him. The Prime Minister's Office advised the Bangladesh High Commission that Mrs. Gandhi would not be able to meet the Deputy Prime Minister forcing them to agree to deliver the letter through the normal diplomatic channels. This treatment of Indian government did not go down well with Bangladesh. At the end of the 5 years period, both parties concluded that water sharing arrangement for the lean season had worked according to the plan, though they could not reach any agreement on the augmentation issue. The final review of 1977 Agreement concluded that the matter of augmentation would be taken up at the higher political level. However, just before that Bangladesh witnessed another coup and General Hussain Muhammad Ershad came to power in Dhaka.

There was no substantial change in General Ershad's foreign policy towards India in general and the Farakka in particular. Nor did Mrs. Gandhi extend a hand of friendship to General Zia's successor in Dhaka. That is perhaps because Mrs. Gandhi was generally adverse to the idea of military regimes in the Indian neighbourhood, abiding by the conventional wisdom that military regimes in South Asia hindered the development of friendly ties between India and its neighbours partly because they sought to divert their respective public's attention away from the domestic problems by constantly looking for a foreign hand (read India) as the source of all their domestic problems and, partly because they suffered from a legitimacy crisis. Experiences with Pakistan had also led Mrs. Gandhi's government to "an easy association of military rule in a neighbouring country assisted by the US with enmity towards India".[36] Mrs. Indira Gandhi's discomfort with the military regimes presence in the neighbourhood is evident from her indirect statement, "there have been nations where democracy was subverted by men of little vision or of overwhelming ambition and such men have been exploited sometimes unwillingly by powerful internal and external forces",[37] shows indictment of such

[36]Surjit (1984), p. 277.

[37]Indira Gandhi (1984), p. 72. Also see, it is quite true that "we do not favour military dictatorships and things like that". Indira Gandhi (1980–1981), p. 552. Also see, Ziring, op. cit., 1978, p. 728.

regimes. Giving description of some of the incidents, the Foreign Secretary of that time writes, “there were, of course, some instances where I was disappointed with Mrs. Gandhi’s rigid attitudes. She was needlessly cool to Admiral (M. H) Khan, the Deputy Chief Martial Law Administrator of Bangladesh when he, along with Abbas called on her prior to the reference of placing the Farakka question in the General Assembly. This meeting between equivalent heads of government barely lasted a few minutes because she observed a stony silence”.[38] As argued earlier, Mrs. Gandhi believed that adoption of democratic and secular principles in the Bangladesh’s constitution had vindicated India’s stand and hence she did not take a benign view of removal of secularism from the constitution of Bangladesh.

General Ershad visited India in October 1982, and the two sides signed an interim agreement for 18 months that covered the dry seasons of 1983 and 1984. The terms of reference were based on a revised sharing formula with marginal differences in allocation albeit with a significant and conspicuous absence of the guarantee clause as was stipulated in the 1977 Agreement.

It is widely assumed that General Ershad was expecting to negotiate a major Agreement on water in order to gain legitimacy in the eyes of its populace but due to Mrs. Gandhi’s aversion towards the military governments, he had to agree only for a Memorandum of Understanding on the water issue. Prof. Asif Nazrul echoes this point that President Ershad negotiated with India from a weaker footing.[39] The MOU had two critical components: the allocation of water and strategies for its augmentation. Discussions on augmentation, once again, proceeded along the well treaded, traditional positions. Bangladesh proposed construction of dams in Nepal to increase availability of water to all parties in the dry season. On the other hand, India proposed a barrage across the Brahmaputra at Jogighopa in Assam combined with a 324 km long link canal across Bangladesh to a point just above Farakka in West Bengal. Along with construction of three dams across north-eastern states of India, as argued earlier, this was envisaged not only to fulfil the requirements of both countries but also to improve navigation, fishing and groundwater supplies.

Each proposal was rejected by the other side. Indian negotiators argued that Bangladesh would be using all available Ganges water but the major river of Brahmaputra would be left running into the sea. Bangladeshis insisted that this was beyond the preview of the Joint River Commission that must focus only on the Ganges river waters. India was not convinced because the latter’s proposal, as before, had not done sufficient homework in terms of the submergence and displacement costs as well as its acceptability to Nepal. India’s Irrigation Minister, Mr. Kedar Pandey sought to dispel Bangladesh’s concerns and draw upon the expert opinion to argue that diversification of sweet waters from the Brahmaputra to the

[38] Mehta (2010).

[39] Personal interview of the author with Prof. Asif Nazul Professor at Dhaka University on 23 October 2009.

Ganges through a link canal would remove salinity of the Ganges, but to no avail.[40] In order to justify its proposal, Bangladesh had presented exaggerated demands of water.[41]

The real bone of contention has not changed over the years. India considers Ganges-Brahmaputra–Meghna as a single system and based on the principle of equitable utilisation; it claims a predominant share of the lean flows of the Ganges river while Bangladesh does not consider the Ganges–Brahmaputra–Meghna to be a single integrated system, often citing Economic Commission for Asia and Far East (ECAFE), the World Meteorological Organisation and the Indian Irrigation Commission (1927) which classifies three rivers as falling within three separate international basins and refuses to countenance any scheme for major inter-basin transfer. Thus, stalemate has persisted on the issue of augmentation.

1.2.6 The Domestic Pressures

General Ershad was facing a difficult situation at home mainly due to his regime's lack of political legitimacy. Emulating the strategy of General Zia ul Haq in Pakistan, Ershad also turned to Islam and proclaimed it to be the state religion and lifted the ban on all religious political parties. His declaration of Islam as state religion gave rise to ISA (Islamic Shansonotantrannadolan) as the religious political party which questioned his legitimacy to make Islam as state religion. Brigadier M. Abdul Hafiz (Retd.) believes that this became another important source of growing differences between India and Bangladesh.[42]

At the home front, this led to a quick and mushrooming growth of smaller parties resulting in multiple, discordant voices on the Farakka issue. The Bangladesh Labour Party criticised the government for its inability to withstand and counter anti-Bangladesh activities of India and claimed that through the Farakka barrage, India was planning to turn eight northern districts of Bangladesh into a desert.[43] Mr. A.N.M. Abdul Hamid, General Secretary, Bangladesh National Awami Party (Bhashani), urged everyone to create a strong public opinion to compel India to give legitimate share of the common rivers to Bangladesh.[44] Maulana Mohammadullah Hafezji Huzoor, Amir-e-Shariat of the Khelafat Andolan, characterised the Farakka barrage as "anti-religious and anti-humanity" and said that the problem was not only political but also religious. Secretary General of Bangladesh Jamaat-e-Islami A.K.M Yusuf said that the people of Bangladesh would resist the

[40]JRC Talks Fail (1980).

[41]Verghese (1999), pp. 362–364.

[42]Personal interview with Brig. M. Abdul Hafiz (Retd.), former Director of BIISS, at Dhaka on 25 October 2009.

[43]"Farakka Resistance Day Observed in Dacca", *BSS*, 16 May 1981.

[44]"Call to Withhold Transit Accord", *Bangladesh Times*, 24 May 1983.

link canal proposal at any cost and called upon the Muslim world to raise its voice against India's violation of international law.[45] Khondaker Moshtaque said that the Farakka was a national problem. Referring to the formation of the Jatiya Oikya Front, he said that it had been formed with the objective of resisting aggression, safeguarding national sovereignty and establishing democracy in the country. Mrs. Amena Begum, Mr. S.M. Sulaiman and Mr. Abu Naser Khan Bhashani of Farakka and Border Attack Resistance Committee, urged the government to withhold the implementation of the agreement between India and Bangladesh on the road transit to protect Bangladesh's sovereignty.[46] All the political parties were competitive in raising the pitch against India and called upon people to boycott Indian goods and stop business transactions with India. Extending its reach beyond the borders, the United Kingdom unit of Farakka Songram Parishad was formed in London,[47] because they had been demanding internationalisation of the Farakka issue at various world bodies including the United Nations and the World Bank.[48] While observing the Farakka day on May 16, the leaders of various political parties called upon the Third World and Muslim countries to extend their support in solving the Farakka problem. This meeting was addressed by the local Bangladesh Nationalist Party, Muslim League, Islamic Democratic League leaders and the unit commanders of the Muktijodha Sangsad. The National Action Committee on the Farakka criticised the government for its silence and inability to take the issue to the United Nations for a just and legitimate solution. They decided to send their own delegations to the United Nations and different countries, especially in the Muslim world to launch an international awareness campaign about India's unfriendly acts against neighbouring Bangladesh.[49] Another group of political parties including Khondaker Moshtaque Ahmed-led National front were lobbying for the restoration of democracy to resolve the Farakka issue. No breakthrough was, however, achieved.

1.2.7 Bilateral Talks: Turning a New Leaf?

After the assassination of Mrs. Indira Gandhi, her son and successor Mr. Rajiv Gandhi came to power and directed the top government officials to re-start the negotiation process. Mr. Ramaswamy Iyer, who was then Water Secretary of the

---

[45]Ibid.

[46]"Khondaker for United Resistance", *Bangladesh Times*, 24 May 1983.

[47]"Farakka Resistance Body Calls For Boycott of Indian Goods", *Bangladesh Times*, 23 May 1983.

[48]"Farakka Resistance Day Observed in Dacca", *BSS* 16 May 1981.

[49]Ibid.

Indian government, reiterated this point that the new Prime Minister made an earnest effort to resolve the long-standing issue of water sharing through political negotiations which had not yet taken off.[50] In the aftermath of May 1985 cyclone devastating some of the southern islands of Bangladesh, Indian Prime Minister Rajiv Gandhi flew to Bangladesh to offer Indian assistance. He met Bangladesh President Ershad and, amongst other topics, they discussed the question of sharing the Ganges. This meeting began a new series of discussions and initiatives. Again Rajiv Gandhi met President Ershad at the Commonwealth Heads of Government Conference at Nassau (Bahamas) in October 1985 and discussed Ganges water among other issues. Later, the MOU was negotiated at Minister-level. An interim "memorandum of understanding" was signed in November 1985. The MOU was modelled after the 1982 understanding. It was decided to approach the issue afresh. A new Joint Committee of Experts alongside the Joint River Commission was created. The JCE was headed by their irrigation secretaries for a 12 months period.

Meanwhile, Bangladeshi Minister for Irrigation, Water Development and Flood Control began exploring resolution of the Ganges issue in a different way. He was guided by the idea that if Bangladesh could have clear idea of its share from common rivers, it could proceed to some long-term alternative plans of the development and utilisation of its water resources accordingly. He proposed informally that Bangladesh could get 25,000 cusecs at the lowest level of the Ganges flows of 55,000 cusecs; in Brahmaputra, India and Bangladesh would share the lean season flows on a 25–75 basis, with the latter retaining 25 % for various uses and allowing 50 % of flow to the sea for preventing saline incursion and for other environmental purposes. In other common rivers, the share of the two countries should be 50:50.[51] When Bangladesh Minister of Irrigation was about to present the new proposal based on Expert Study Group to the cabinet chaired by General Ershad, meeting was interrupted and the proposal was never discussed again.[52]

Bangladesh witnessed massive flood in 1988 resulting in submergence of at least two-thirds of its territory. The top leaders cutting across the political spectrum held the Farakka responsible for the deluge.[53] President Ershad reiterated in numerous public speeches that waters causing so much death and destruction in the country flowed from beyond the borders and the damage might have been mitigated if there had been the right mix of concern and commitment shown by India.[54] The Indian Air Force's rescue and relief mission in Dhaka was suddenly told that it was no

---

[50] Personal interview with the author at New Delhi on 26 December 2008.

[51] Rangachari and Iyer, op. cit., pp. 190–191.

[52] Crow with Alan Lindquist and David Wilson, "The Rise and Fall of New Initiatives: 1983–1987", in op. cit., pp. 206–207.

[53] There has been tendency in Bangladesh to blame Farakka both for reduced flows and for floods. India has always conveyed to Bangladesh that Farakka is a barrage and not a dam, that it is a diversion and not a storage project, and that while it can cause a reduction of flows, it cannot possibly cause floods.

[54] Hasan, (1991) p. 35.

longer required which in turn, evoked strong reactions from India. Different political parties competitively accused India for causing such devastation. The Acting Ameer of Jamaat-e-Islami, Mr. Abbas Ali Khan, for instance, blamed the subservient foreign policy of Ershad regime and demanded his resignation for its failure to find a way for permanent solution of the Farakka problem.[55] The Bangladesh Nationalist Party (BNP) Chairman, Begum Khaleda Zia observed that the devastating floods were mainly due to construction of various barrages including Farakka in the upstream of many Indian rivers.[56] Even politically fringe elements like Maulana Matin took advantage of the situation and launched a boycott of Indian goods in a bid to hijack the Farakka agitation platform against India".[57] The real problem lay in the lack of legitimacy and stability of the Ershad regime because that, Reaz Rahman explained to this author, "pushed Ershad towards adhocism".

At the same time, when Ershad was trying to mollify the domestic opinion, Mr. Anishul Islam Mahmud, the irrigation minister in the Ershad regime in 1984, told this author about his efforts in persuading and convincing the press, members of parliament that separating the issue of water sharing from augmentation was in Bangladesh's interest.[58] The groundwork for signing the treaty was, in fact, done at Malaysia where the Indian Prime Minister Rajiv Gandhi and President Ershad were scheduled to meet in 1989. At this time, the former announced general elections resulting in yet another regime change in New Delhi and Mr. Ershad too had to succumb to pressures of domestic agitation for restoration of democratic government; an election was held in 1991 which brought Mrs. Khaleda Zia to power. The Bangladesh's constitution was changed from the Presidential system to the Parliamentary form of Government and Mrs. Khaleda Zia then became the Prime Minister of Bangladesh.

1.2.8 Farakka: The Political 'Football' in Bangladesh's Domestic Politics

The India–Bangladesh's negotiations on Farakka are marked by a paradox. Interestingly, the military regime of Ershad could have single-handedly negotiated a water treaty and was keen to do so as an instrumentality to gain legitimacy for his regime. New Delhi did not oblige precisely due to that reason. Mrs. Indira Gandhi believed that signing a water treaty with a military regime would bestow legitimacy

[55]"Farakka Cause of Bangla Flood, Alleges Jamaat Leader", *Bangladesh Times*, 7 September 1988.

[56]"Khaleda Asks India to Help in Solving Problem Permanently", *New Nation*, 22 September 1988.

[57]"Political lumpens to the Fore", *Holiday*, 28 May 1983.

[58]Personal interview with author at Dhaka on 25 October, 2009.

upon him and weaken the democratic forces in Bangladesh which, in her view, was more important for strengthening the long-term relationship between the two countries. However, the onset of democracy in Bangladesh heralded an era of competitive politics, in which the Farakka was used as a political 'football' between its two mainstream political parties. A concession, in fact, *any* concession made by one party was debunked as a 'sell-out' by the other sitting on the opposition benches, Awami League (and later, BNP, in turn), missed no opportunity of accusing the ruling party of making Bangladesh subservient to India, thus rendering it practically impossible to arrive at a bipartisan consensus for negotiating any mutually acceptable agreement with New Delhi. Paradoxically, the tables would turn every time in the sense that every party that won elections and came to power was forced to face the harsh reality that without adopting a conciliatory and accommodative approach of mutual "give and take", negotiations with India would simply not yield any positive results. However, by then, its hands would be tied by those sitting on the opposition benches clamouring for a hard-line approach. This cycle continues till date. The following section explains this phenomenon which got underway in the period leading to 1991 elections.

The end of martial law in 1990 paved the way for elections to be held in Bangladesh in 1991. The issue of the Farakka and the larger Indo-Bangladesh relationship was a key point of contestation among different political parties. The BNP, led by Begum Khaleda Zia attacked the Awami League fiercely for being sympathetic to India and issued a dire warning that if BNP did not come to power, the country would be sold by the Awami League and the capital of Bangladesh would be New Delhi instead of Dhaka.[59]

After coming to power, however, Khaleda Zia tried to normalise relations with India; she visited the neighbour in 1992 and signed an eleven point Joint Statement which mentioned joint efforts for reaching an equitable, long-term and comprehensive arrangement for sharing the flows of major rivers. This evoked a huge hue and cry back home because the opposition parties had been clamouring for more water. The Awami League's leader Sheikh Hasina termed it as "treaty of bondage" and demanded it be scrapped. Countering the attack against her party by BNP during electioneering, it was Sheikh Hasina's turn to accuse the government of having "masters, no foreign friends".[60] Since Bangladesh already had a benchmark in terms of getting water allocations from India, the BNP government was brought under the tremendous pressure to secure at least that much, if not more. Sheikh Hasina accused that while the Awami League's government had realised 44,000 cusecs of Ganges water during the dry season in 1974, the Zia ur Rahman government had got 34,500 cusecs, 10,000 cusecs less than the Awami League government and now Begum Zia had failed to fetch "even a mug of water" from New Delhi.[61] She challenged the government to "prove that you are not serving the

[59] "River Waters Issue Being Used for India-Batting", *Sunday Express*, 15–21 September 1991.

[60] "Hasina Demands Scrapping of Treaty with India", *Morning Post*, 25 October 1993.

[61] "Hasina says Khaleda sold out Bangladesh to India", *Morning Sun*, 12 November.

interest of India by not realising the just share of the Ganges waters".[62] In such circumstances, Begum Zia became hesitant to pursue further water negotiation with India until she was sure of receiving the same volume of water as demanded by the opposition. And that did not seem likely because as Prof. Asif Nazrul pointed out, Dhaka was aware that sufficient water was simply not available in the Farakka at that time.[63]

Other smaller political parties had also joined the bandwagon in criticising Begum Zia's government. The Muslim League President Kazi Abdul Kader demanded immediate cancellation of 25 years treaty and pledged that his party would fight unto death in realising due share of the Ganges water.[64] Surprisingly even the Awami League Joint secretary Amir Hussain Amu called for striking out all clauses detrimental to the interest of Bangladesh in the 25 year Indo-Bangladesh Friendship Treaty which Mujibur Rahman had signed when he first came to power.[65] The Democratic League argued in a similar vein that the Farakka problem could not be solved due mainly to the subservient foreign policy of the BNP government and, the Freedom Party termed the treaty as "regrettable, condemnable and suicidal".[66]

The agitation over the Ganges was not confined only to political parties but other sections of society were also involved. The Dhaka University students had launched an action programme to press for immediate settlement of the Farakka dispute and cancellation of the 25-year friendship treaty between India and Bangladesh.[67] The Bangladesh Environmental Lawyers Association, Secretary General, Dr. Mohiuddin Farooque argued in favour of lobbying around the world for getting equitable share of the Ganges water and emphasised third party mediation. The leaders of the Farakka-O-Aggression *Protirodh* Council had called upon the government, opposition political parties and professional bodies to become united on a common platform for securing compensation from India by filing a suit in the International Court of Justice. The Council also announced an action programme including submission of memorandum to foreign missions, mass contact, rallies and a national convention at Dhaka, a grand rally in the capital and Farakka long-day march on May 10, 1995.[68] Various organisations in Bangladesh had been observing the Farakka Long March Day on May 16 every year. This march was started by Maulana Abdul Hamid Khan Bhashani in 1976 who had led a long march from Rajshahi town towards the Farakka on this day to raise this demand. Terming the

---

[62]"Hasina Asks Govt to secure Due Ganges Water Share", *The Daily Star*, 8 January 1994.

[63]Personal interview with the author at Dhaka on 23 October 2009.

[64]"ML For Filing Compensation Case Against India on Farakka", *Bangladesh Observer*, 13 January 1994.

[65]"Indo-Bangla Friendship Treaty Imbalanced, Says AL Secy", *Morning Sun* 3 April 1991.

[66]"Freedom Party Slates Govt", *Bangladesh Observer* 19 May.

[67]"Cancel Indo-Bangla Friendship Treaty: Demands DUCSU Meet", *Morning Sun*, 2 October 1990.

[68]"Demand for Compensación From India", *Bangladesh Observer,* 30 May 1994.

Farakka as a political weapon, JAGPA leader Shafiul alam Prodhan said the Indian decision to construct the barrage was a ploy designed to weaken its neighbour and urged the government to approach the International Court against India asking for compensation for the economic damage done to Bangladesh by the Farakka barrage.[69]

Though the BNP government did not take this radical step, it adopted a twin-track strategy with regard to the Farakka issue like the Zia government. First, as explained earlier, it explored the bilateral track of negotiations with New Delhi to find a mutually acceptable settlement and, second, it sought to internationalise the issue in various international fora. First did not make much headway partly due to technical constraints emanating from the lack of availability of sufficient water to satisfy demands made by both sides and partly, and more importantly, due to their political differences over other issues such as illegal migration of Bangladeshis into India. The former foreign Secretary, Mr. Reaz Rahman, who had accompanied Khaleda Zia for her meeting with the Indian Prime Minister, P.V. Narshimha Rao, explains this point. In an obvious reference to the illegal Bangladeshi migrants, Prime Minister Rao told Mrs. Khaleda, "Madam, if you go to Gujarat and address the rally in your language there, at least 25,000 people would come to listen to you", and Mrs. Khaleda replied "Mr. Rao, you are confused with your Bengali people. They speak the same language but it does not mean that that they are Bangladeshis".[70] This incident further soured the bilateral relationship between the two countries. It is important to note that the Indian Prime Minister P.V. Narshimha Rao's return visit to Dhaka did not materialise because just 1 month before that, the Bangladesh foreign Minister Mustafizur Rahman had called Dhaka-based foreign diplomats to see the desertification and ecological degradation caused by India's unilateral withdrawal of common waters in January. In fact, the Indian Prime Minister's visit to Dhaka did not take place till the end of Begum Zia' regime. Though India and Bangladesh's foreign secretaries, water resource ministers and Joint Rivers Commissions kept meeting at regular intervals, the latter missed no opportunity to raise it at different international fora. Begum Zia's effort to internationalise the Farakka issue is discussed in third chapter.

In such circumstances, it became well-nigh impossible to expect a long-term water negotiation which the then BNP government had been demanding. Leave aside a long-term water-sharing treaty, even a MOU could not be signed between the two countries. The Bangladesh Times editorial summed up the situation:

> The water sharing issue is not merely technical and bureaucratic but also largely political. And because it is so overwhelmingly political a solution has so far evaded us. The necessary thrust on the experts will have to come from the ministers and the political leadership of the two countries. Bangladesh hopes that there will be no lacking of *political thrust from India*.[71]

---

[69]"Farakka March Day Observed: Just Water Share Demanded", *Bangladesh Observer*, 17 May.

[70]Personal interview with the author at Dhaka on 26 October 2009.

[71]"Political Thrust to Resolving Water Sharing", *Bangladesh Times*, 29 August 1992.

Interestingly towards the end of the Khaleda Zia's government's term, she once again, adopted an anti-India stance and cautioned people that the coming elections would determine the future of Bangladesh's independence and sovereignty and that if the BNP was not elected, "the country might again be shackled by foreign powers".[72] The Awami League President, Sheikh Hasina however, quickly replied back that in a reference to the 25 years of the Indo-Bangla Friendship Treaty, she challenged Khaleda Zia that: "If you call it a 'treaty of slavery', why do not scrap it?"[73] Sheikh Hasina alleged that the BNP did not fulfil any of its pre-election promises and though it had raised anti-Indian slogans before election, but after capturing power, "it completely handed over Bangladesh to India in the last 4 years"[74] and that the entire northern part of the country had turned into a desert because of Farakka, but the Prime Minister forgot to raise the issue during her visit to New Delhi and rather signed an accord "selling out the country to India".[75]

1.2.9 Fifth Phase: The Final Breakthrough

In 1996, new governments came to power in both New Delhi and Dhaka. In Bangladesh, the Awami League government headed by Sheikh Hasina won the election, while in India a non-Congress coalition government led by Deve Gowda came to power. One of its important constituents was the Left Front, which was in power in West Bengal at that time. Mr. I.K. Gujral became the Foreign Minister of India.

The circumstances in mid-1996 seemed extraordinarily propitious to adopting a fresh approach towards the issue of water sharing mainly because both New Delhi and Dhaka, in a rare display of political resolve, agreed to take the negotiations undertaken thus far to its logical conclusion and conclude a deal on water sharing.[76] The Awami League had returned to power after 20 years-long gap and New Delhi was keen to strengthen this regime in Dhaka. Both realised the need to overcome the impasse over the Ganges waters issue, and showed the urgency to arrive at an agreement before the onset of the next dry season. Farooq Sobhan, the then foreign Secretary of Bangladesh in 1996, shared this assessment and noted that "Awami League had stronger commitment towards having treaty. India also reciprocated in the same manner. Traditionally BNP has been less receptive to India".[77] This was further facilitated by a long overdue realisation on both sides that "an impasse over

[72]"PM warns of AL Again Selling Country's Sovereignty", *Morning Sun*, 21 June 1995.

[73]"Water Sharing Can be Solved By An Al Govt: Hasina", *Morning Sun*, 25 May 1995.

[74]"Hasina Accuses Govt of Selling Bangladesh to India", *New Nation,* 21 March 1995.

[75]Ibid.

[76]For details on this issue, see Iyer R(2003), pp. 236–238.

[77]Personal interview with the author on 26 October 2009.

the Ganges waters issue was unfortunate and that it was desirable to arrive at an early understanding on water sharing putting aside the augmentation question".[78]

The ground work for this was, indeed, laid during the Khaleda Zia's regime, which can be traced back to the SAARC Summit at New Delhi in May 1995 when the then Prime Ministers of India and Bangladesh had, during their bilateral meetings, asked their respective Foreign Secretaries to try and break the logjam by holding talks. One round of talks was held immediately thereafter and they agreed to follow these up later at Dhaka. Foreign Secretary of India, Mr. Salman Haidar, visited Bangladesh during June 23–25, 1995 and the two sides reached an agreement on the following principles: first was to arrive at a permanent sharing arrangement on the basis of existing dry season flow in the Ganges without linking it to the augmentation question. This marked a very significant departure from the previous Indian position. Second was to revive the Joint River Commission expeditiously to work out the modalities for water sharing. JRC, among other things, should exchange information on other projects like Tipaimukh, and so on. Thirdly, they agreed to jointly monitor the flow of the Ganges at selected points as also to arrive at a sharing arrangement of other common rivers on long-term predicable basis. However, no further progress was made in the wake of the political uncertainties in both countries and a period of stalemate ensued with practically no high level contacts between the two countries.[79] These were resumed only after Sheikh Hasina came to power in 1996.

The Bangladesh Foreign Secretary visited India during August 6–10, 1996 to prepare the ground work.[80] During this visit, the Foreign Secretary also held a meeting, at the suggestion of the India's External Affairs Minister, I.K. Gujral, with West Bengal Chief Minister to seek his help and support in finding a permanent solution to the problem of sharing of the Ganges waters. Mr. Basu assured that he would do everything possible on his part to bring the matter to an amicable resolution. Mr. Verghese believes that "involving the Chief Minister of West Bengal was a critical departure from previous negotiation practices".[81]

During September 1996, the Indian External Affairs Minister visited Bangladesh and the two sides agreed to convene a Joint Committee involving members of their respective foreign ministries which were tasked to bring water experts from both sides to work towards finalisation of an agreement on water sharing. The formation of such a committee was another major departure from the past where such committees had always been under the jurisdiction of the Water Resources Ministry. By bringing the technical experts under the supervision of the political leadership the discussions were "moderated in a manner so as to lend flexibility and give

[78]Iyer (2003), p. 235.

[79]Tariq (1998), pp. 225–226.

[80]Ibid., p. 227.

[81]Ibid.

pre-eminence to the political agenda over the obfuscation of engineering technicalities".[82]

The Bangladesh Minister for Water Resources' visit to India from October 28 to November 1, 1996 to hold further talks with his Indian counterpart was quickly followed by that of the Foreign Minister's visit and the meeting between the Joint Committee of Experts. During November 9–13, 1996, both sides reaffirmed their commitment of arriving at an agreement on a fair and equitable sharing of the Ganges waters before the onset of the next dry season. The Bangladesh Foreign Minister also visited Calcutta to get Mr. Jyoti Basu actively involved in the negotiating process. In fact, Mr. Basu's visit to Bangladesh from November 27 to December 2, 1996 proved to be very crucial because it helped significantly narrowing the differences between the two sides. He had wide-ranging discussions with both the Bangladesh's Prime Minister and the Foreign Minister, while the West Bengal's Finance Minister and other senior officials of his government and the central government held intensive parleys with the Bangladesh's Joint Committee members led by the Foreign Secretary and subsequently, the secretary to the Prime Minster of Bangladesh. While the negotiations were underway at the political level, the Joint Committee met several times to re-examine the technicalities involved and worked hard towards arriving at an agreed draft agreement to be placed before the two governments. This culminated in the last marathon session of the Committee from December 5–10, at New Delhi at the foreign secretaries' level. It was during this final round of negotiations that the nature of the instrument to be drawn up—Treaty instead of Agreement—and the duration of 30 years were agreed upon.

The final product, the treaty on the sharing of the Ganges waters entered into by India and Bangladesh on December 12, 1996 was "a more significant document than most people had considered possible".[83] Iyer correctly sums up the driving factors that led to signing of this treaty:

> The patient efforts of the Indian Ministry of External Affairs from 1995 onwards to find a solution to this problem; the wisdom and courage of the Prime Minister of Bangladesh, Sheikh Hasina, in tackling an issue fraught with considerable political risks; the high priority that the new Indian Foreign Minister attached to the resolution of the Ganges water dispute; the constructive and sagacious role played by the Chief Minister of West Bengal and the contributions made by his Finance Minister towards finding answers to the difficult questions that came up in the negotiations; and last but not least, facilitatory efforts at non-official levels. It must be added that quite understandably and necessarily, *it was a political rather than a technocratic settlement.*[84] (italics added)

It is important to highlight some of the important features of the Ganges treaty which are as follows:

a. India will release water from the Farakka barrage in the 5 month lean season, from January 1 to May 31, every year;

[82]Ibid., p. 228.

[83]Iyer (2003), p. 236.

[84]Ibid., pp. 235–236.

b. Bangladesh will get a minimum of 35,000 cusecs, or 50 % of the Ganges water at Farakka, if its volume is 70,000 cusecs, or less;
c. India and Bangladesh will get guaranteed 35,000 cusecs in three alternative 10-day periods from March 1 to May 10;
d. Bangladesh will get a maximum of 67,516 cusecs from January 1 to 10, and a minimum of 27,633 cusecs from April 11 to 30
e. India will get a maximum of 40,000 cusecs in seven 10-day periods in January and February and from May 21 to 31 and a minimum of 25,992 cusecs from April 21 to 30 and
f. if the flow at Farakka falls below 50,000 cusecs in any 10-day period, the two governments will discuss adjustments.

Soon after the treaty, a survey was conducted in Bangladesh to assess the reaction of Bangladeshis about the treaty, which reported that 89 % per cent of the respondents believed that it would be beneficial for Bangladesh. The BNP protested against the treaty and claimed that Bangladesh's national interests had been compromised and insisted on incorporating a guarantee clause in the Treaty. The Bangladesh Muslim League (BML) argued in a similar vein that the interests of the country were not protected in the treaty because there was no guarantee clause; Nepal was not included, and there was no provision for the presence of any third party.[85] However, the majority of population did not share this assessment. The survey reported that people observed that the Awami League government had not betrayed the national interest by signing the treaty. More than 84 % people believed that the increased flow of water under the treaty would have a positive impact on the national economy.[86]

1.2.10 A Critical Evaluation of the Functioning of the Treaty

The 1996 Treaty has three parts: the preamble, the operative part containing twelve Articles and the Annexes.[87] Article I to XI set forth the provisions for sharing the Ganges flow and related matters. Article XII puts the duration of the treaty for 30 years and it could be renewed on the basis of mutual consent. Though the 1977 agreement contained a "guarantee clause" and the 1985 deal had a "burden sharing" formula, the 1996 Treaty did not include any compulsory in-built safeguards for Bangladesh. There are however various provisions which provide a modicum of

[85]"India can make water treaty ineffective: Muslim League", *New Nation*, 6 January 1997.

[86]"96 % Consider Water Treaty Essential: Opinion Poll", *Bangladesh Observer*, 25 December 1997.

[87]See Annexure 3 for a complete text of The Treaty between the Government of the People's Republic of Bangladesh and the Government of the Republic of India on sharing of the Ganges/Ganges waters at Farakka.

security. The treaty recognises that the flow in the river may reduce over a period of time and thus, it requires the government of India to make every effort to protect the flows arriving at Farakka. It is written in clause (ii) of the Article II of treaty that "every effort would be made by upper riparian to protect flows of water at the Farakka as in the 40 years average availability". Another important aspect is that when the flow goes below 50,000 cusecs, the Treaty recognises an emergency situation and provides for immediate consultations by the two governments. There is, however, a provision for a review at the end of 5 years or even at the end of 2 years if either party so desires. The Treaty lays down that by keeping pending agreed adjustment as a result of such a review, India will release to Bangladesh not less than 90 % of its entitlement under the Treaty. It also provides for a conflict resolution mechanism by prescribing a joint monitoring of flows, which should eliminate or minimise the possibility of disagreements over the data. If the Joint Committee is unable to the solve the dispute, the matter is then referred to the Joint Review Committee (JRC), and failing resolution at that level, the issue will be taken by the two governments.

In 1997, three issues emerged regarding the Ganges treaty. The issues were low flows, alternate 10-days pattern and Farakka–Hardinge Bridge discrepancy. The treaty faced its first test only a few months after it came into force because the actual availability of the waters of the Ganges at Farakka turned out to be far less than the average flow of the Ganges during the period from 1949 to 1988. The Indian side reiterated that 1997 could be considered as an abnormal year. Even the historical data showed that before the Farakka Barrage was built, flows had fallen as low as 40,000 and 39,000 cusecs in 1952 and 1953 respectively. Usually, the low point in the flow occurs between the second 10 days of March to the first 10-days of April. But the turning point in 1 year might be as early as the first 10-days of March as well as in April. The most notable feature of the 1977 Agreement was that it included a clause that guaranteed Bangladesh a minimum of 80 % of its share of water during any 10-day period, irrespective of how low the flow of the Ganges River might be during such a period. Farooq Sobhan explains that according to the 1977 Agreement, the river flow would be exceptionally low if it was less than 80 % of the then 75 % dependable flow of 55,000 cusecs or 40,000 cusecs. The flow had actually gone down to 48,487 cusecs in 1980 and 43, 555 cusecs in 1983".[88]

There was another issue pertaining to the discrepancy between the quantum of water released at the Farakka barrage in India and that arriving at the Hardinge Bridge—170 km downstream—in Bangladesh, which became a major bone of contention between the two countries in 1997. Several explanations have been offered to describe such a discrepancy. At Hardinge Bridge, the Ganges channels are very large and have a carrying capacity of 1.5 million cusecs, but bad load movement, sediment distribution and sand bar formation can be peculiar, rendering it difficult to measure the flow correctly. Alternately, it could be due to some developments between the Farakka and the Hardinge Bridge specially relating to

[88]Sobhan (2002), p. 63.

the major tributary of the Mahananda and Bhagirathi that lie in between and are dry while the Mahananda no longer contributes any inflows because of upstream withdrawals. Another assumption is that the river water might be going into underground aquifers in Bangladesh.[89] The problem might have got exacerbated due to low flow in 1997.

Another issue came to the fore regarding the operation of the treaty was the alternate sharing of 35,000 cusecs to either side in alternate 10-day periods between March 11 and May 10. This leads to sharp variation in flows from one 10-day period to the another one. This provision has got to do with safety issue especially when the flows reaching Farakka are low. A drastic drop in the water level in the feeder canal on the Indian side, if carried out rapidly, could lead to a collapse of the canal sides. A gradual reduction in the canal level in the interest of safety resulted in a shortfall in the releases to Bangladesh in the first 10 days of April which was compensated within the next 10-day period. This further led to charges of violations of the treaty.[90] Verghese explains, this is "technically unsafe, as the walls of the Farakka feeder canal could collapse with such abrupt changes in levels".[91] He further explains that since it is not a system where you can switch on or off according to 10-day pattern, water flows increase gradually and the same pattern is followed towards the end of the 10-day period but water does not stop on the tenth day. He underlines the importance of dispelling mistaken notions among some Bangladeshi people who assume "somebody is sitting there whose job is to switch off or switch on the key of water flows" and that this needs to be understood by people across the borders.

Both countries have accepted switching of 10 day flows to one side or the other during six alternating 10-day periods due to their experience from the previous negotiations. That is why, a stepped pattern was written into the treaty instead of the smoother gradient spread over a few days whereby the quantum of water delivered would have remained the same though with a different 10-day flow pattern.[92] The Indian side had no option but to stagger the reduction and augmentation of releases over some days, which altered the pattern of scheduled deliveries in a few 10-day periods. Bangladesh was compensated for any scheduled deficit in supplies in the ensuing period. This was cited as a treaty violation, though actually the total quantum of water released to Bangladesh over the lean season as a whole was a little more than stipulated in the schedule.[93]

The natural conditions also play an important role in the maintenance of water flows in the Ganges which mainly depends on rainfall and snow melting in the upstream areas of the river. Records of water flows at different stages in the Ganges river as well as the data of the North Indian rainfall stations and normal snow melt

---

[89]Iyer (2003), p. 241.

[90]Ibid.

[91]Verghese (2001), p. 173.

[92]Ibid.

[93]Ibid.

in various locations in Nepal and a comparison of this data with the data of a particular year, can make it possible to estimate the effect of precipitation and snow melting in the river flow. It will, however, be difficult to precisely estimate the contribution of snowmelt to the flow at the Farakka because of long distance travelled by Ganges between the snow melt in the Himalayas before it reaches at the Farakka. In other words, the issue of discrepancy is a complex matter which needs a scientific study since "it had not been explained to the people", they were therefore ready to be persuaded by the opponents of the treaty that they were being short-changed at Farakka",[94] unnecessarily fuelling the matter out of proportion.

There is yet another and more complex problem of the Gorai hump. Bangladesh's grievance about diversions by India from the Ganges at Farakka has revolved around the acute distress said to have been caused in the northern region on account of salinity ingress and a shortage of water for agriculture, fisheries, navigation, and sustenance of the Sundari mangrove species. This area, because of the Gorai spill, which delivers upland fresh water supplies to the region, is left high and dry as the Ganges recedes. While this is so, it would be erroneous to attribute the problem exclusively or mainly to diversions at the Farakka. The entire Ganges system has been shifting east and north as a secular trend over the past century and more. The Bhagirathi, the westernmost spill, was the first casualty. Other streams, moving further east, have progressively deteriorated as the Ganges has shifted course. This has resulted in the closure of the Gorai mouth by a hard 5-m silt plug that is now said to measure almost 30 km from outfall point to Gorai bridge.[95]

In any case, the Treaty is unlikely to solve this problem, because even 35,000 cusecs are not enough for this purpose; only a water flow to the order of 70,000 cusecs would help the waters of the Ganges in the Gorai river. The answer to this problem is perhaps partly extensive dredging and partly arrangements to head up the Ganges waters and enable them to enter the Gorai.[96] Actually the issue of the Gorai River was not a new one. "In 1971–1972 the World Bank had made a study of the Gorai region and reported that even in the pre-Farakka years, the flow of the Gorai had been low due to the emergence of sand bars for which large scale dredging was required. This had been considered by the government of Bangladesh and President Ziaur Rahman had, at one stage, "initiated labour-intensive excavation activities under the Food for Work Programme which had generated some positive results".[97]

Since the flow of the Ganges river continued to be below 50,000 cusecs, Bangladesh asked India for "immediate consultation" to make adjustments on an

[94]Iyer (2003), p. 242.
[95]Verghese (2001), p. 175.
[96]Iyer (2003), p. 242.
[97]Sobhan (2002), pp. 75–76.

emergency basis as stipulated under Article 11 (iii) of the Treaty. India agreed to hold immediate consultation with Bangladesh, and a series of meetings were held in Dhaka and New Delhi. During the meetings, Bangladesh demanded that India ensure that it would receive the 35,000 cusecs guaranteed under Annexure of the Treaty, and also that India communicate the steps it was taking to protect the flow of water at Farakkka, as per article 11 (ii) of the Treaty. India confirmed that the flow at Farakka had slowed down and attributed this situation to the normal hydrological cycle that occurs every 4–5 years and stated that it was complying with its obligations under the treaty by agreeing to immediate consultation. In mid-June, an expert-level meeting of the Indo-Bangladesh Joint Rivers Commission was held in Dhaka. After 3 days of discussions that centred on the Treaty, the commission recommended that a scientific committee be formed to study the causes of the low flow of the Ganges during the critical period of the dry season. Later, the 30-s meeting of the Joints River Commission was held on July 18–20, 1997. The two sides appreciated the need to remove the bottlenecks in implementation of the Ganges Treaty with further negotiations. It was decided to undertake joint scientific studies in accordance with the terms of reference, which have been jointly finalised by the technical terms of the two countries.

By early August 1997, the flood information centre in Bangladesh had started warning that the Ganges water flow was above the danger mark, and that certain areas could soon be flooded. The 1998 dry season flows helped to ameliorate the concerns raised in the preceding year as the Ganges flow in that year had fulfilled Bangladesh expectations. During the first 10 days cycle of 1998 lean season, the discharge of water at the Farakka and its availability at the Hardinge Bridge point had doubled as compared to the level of January 1–20, 1997. During this monsoon period, the flow created devastating floods. The 1999 dry season flow was less voluminous than that of 1998, but was still far more than what was prescribed under the Treaty. Only on one occasion, Bangladesh objected that it was not receiving water according to the schedule of the treaty. The Joint River Commission stated in a press release issued in Dhaka on April 6,1999 that in one of the six schedules, Bangladesh received more than its share of the Ganges water during the lean period, the latest quantum being 33,892 cusecs at Hardinge Bridge during March 21–31 against 29,688 cusecs as stipulated in the Treaty.[98] The flow of the Ganges during the dry season of the year 2000 was similar to that of the last year though in an important departure, the last 10-day period of April witnessed an increased flow as compared to that specified in the Treaty.

But there have been criticisms against the 1996 agreement as it has been based on the flow average of 1949–1988, but the real flow at Farakka in the 1990s has been much less than that. "To get a reliable figure, the water experts should have

[98]"Ganges Water Treaty: Bangladesh Gets due Share", *The Daily Star*, 7 April 1999.

taken the average of the flow of last 10 years to the agreement. The long-term effect of faster glacial melting is going to be critical for the dry season flow of the river and that will pose serious problems for the bilateral agreement to reallocate the decreasing water in the face of increasing demand".[99] The editorial of the Statesman commented on the 1996 Ganges river water Treaty this way; "It must be said that facts and figures regarding water availability at Farakka have been played around like the West Bengal Finance Ministers' by now infamous Zero deficit budget. This agreement shows average total water flows at Farakka which are not there".[100] Thus, the flow of Ganges keeps fluctuating from one 10-day to another 10-day period but important point is that if overall flow from Farakka to Bangladesh has been maintained or compensated in next 10-day period, this issue should be sorted out at the Joint Committees levels which are stationed at the Farakka and Hardinge Bridge. Only issues of grave concern should go to higher echelons of authorities.

1.3 Conclusion

Thus the discussion on the Ganges has been dominated by politicisation of the issue in both countries. The technical nature of the problem in terms of deciding the quantum of each party's share of common waters and, finding ways of augmenting the Ganges flows have not radically changed over the decades. The only thing which has changed is domestic political discourse. The successful outcome hinges on the art of making the agreement politically palatable. In 1996, when the Ganges treaty was signed, augmentation issue was segregated from water sharing as augmentation involves technicalities. The technical experts have been coming to the same conclusion on means of augmentation in almost all meetings which ensued deadlock. As observed in all three agreements of 1975, 1977, 1996, the strong desires of politicians made sure that obstacles should be circumscribed. Thus, even when the Joint Committee was formed in 1996 to solve water issue, water experts were brought under political control. Thus the 1996 treaty was a political decision and even quantum of water agreed between the two countries was 35,000 cusecs, little more than what 1977 Agreement had agreed to share, making it politically palatable for the ruling party in Bangladesh.

The long-standing conflict on Farakka and tortuous negotiations for resolution of the dispute for decades underlines the importance of analysing water conflicts in the social, political and historical context. These different variables are being explained in the next chapter. These variables bear strong influence on the negotiating approach and outcome of deliberations and further help in explaining how the Ganges river issue turned into major dispute between India and Bangladesh.

---

[99]Swain (2012), p.72–87.

[100]"A Positive Step", *The Statesman*, 14 December 1996.

Annexure

Joint Declaration Issued at the End of the Visit of the Bangladesh Prime Minister Sheikh Mujibur Rahman to India, New Delhi, May 16, 1974

1. At the invitation of Her Excellency Shrimati Indira Gandhi, Prime Minister of India, His Excellency Bangabandhu Sheikh Mujibur Rahman, Prime Minister of Bangladesh, visited New Delhi from May 12 to 16, 1974. The Bangladesh Prime Minister was accompanied by Mr. Khandker Mushtaq Ahmed, Minister of Commerce and Foreign Trade. Dr. Kamal Hossain, Foreign Minister and Dr. Nurul Islam Deputy Chairman, Planning Commission and senior officials of the Government of Bangladesh.
2. During his stay in New Delhi the Bangladesh Prime Minister called on the President of India, His Excellency, Shri V. V. Gin, and had a very friendly exchange of views.
3. The Prime Minister of Bangladesh paid homage to the memories of Mahatma Gandhi and Pandit Jawaharlal Nehru and laid wreaths at Rajghat and Shantivana.
4. The two Prime Ministers discussed matters of mutual interest on several occasions during the visit. These discussions were held in an atmosphere of utmost cordiality and friendship and were marked by frankness and mutual trust, reflecting the strong bonds of friendship and understanding between the Governments and the peoples of the two countries.
5. The Bangladesh Minister of Commerce and Foreign Trade held separate talks with the Minister for Commerce of India.
6. The Bangladesh Foreign Minister also held talks with the Indian Foreign Minister on bilateral relations and on the international situation.
7. The Deputy Chairman of the Bangladesh Planning Commission held talks with members of Planning Commission of India.
8. The opportunity was also availed of to have discussions at the official's level on various aspects of relations between the two countries.
9. Participating in the talks from the Bangladesh side were Mr. Khandker Mushtaq Ahmed, Minister for Commerce and Foreign Trade, Dr. Kamal Hossain, Foreign Minister, Dr. Nurul Islam, Deputy Chairman, Planning Commission, Dr. A.R. Mallick, Bangladesh High Commissioner in India, Prof. Rahman Sobhan, Member, Planning Commission, Mr. B.M. Abbas, Adviser to the Prime Minister, Mr. Nurul Islam, Secretary, Foreign Trade Dr. M.A. Sattar, Secretary to the Prime Minister, Mr. Fakhruddin Ahmed, Acting Foreign Secretary, Mr. Asafuddullah, Acting Secretary, Water Resources Division and other senior officials. On the Indian side were Sardar Swaran Singh, Minister of External Affairs, Prof. D.P Chattopadhyaya, Minister of Commerce, Shri K.C. Pant, Minister in the Ministry of Irrigation and Power, Shri Surenderpal Singh,

Minister of State, External Affairs, Prof. S. Chakravarti, Member, Planning Commission Shri S.Dutt, High Commissioner for India in Bangladesh, Shri Kewal Singh, Foreign Secretary, Shri P.N. Dhar, Secretary to the Prime Minister, Shri Y.T. Shah, Commerce Secretary, Shri R.V. Subramaniam, Secretary, Ministry of Irrigation and Power, Shri BK. Sanyal, Secretary, Ministry of External Affairs and other senior officials.

10. The two Prime Ministers had detailed discussions on 11w situation in the sub-continent and on international developments. The Prime Minister of Bangladesh paid high tribute to the personal contribution of Shrimati Indira Gandhi to the liberation struggle of Bangladesh and her outstanding leadership in the promotion of durable peace and harmonious relations in the sub-continent. The Prime Minister of India stated that but for the high statesmanship and magnanimity shown by Bangabandhu Sheikh Mujibur Rahman it would not have been possible to arrive at the Delhi Agreement of August, 1973 and the recent tripartite agreement between Bangladesh, India and Pakistan.
11. The Prime Minister of Ranlai1esh and the Prime Minister of India reviewed the prospects for normalisation in the sub-continent following the singing of the Tripartite Agreement of April 9, 1974. They agreed that the spirit of mutual accommodation and the reconciliation shown by the Governments of the three countries, which had led to the successful conclusion of this Agreement, should not be allowed to be dissipated and the improved prospect for establishing durable peace in the sub-continent should be further strengthened by concrete action. They expressed the hope that the three countries of the Sub-continent would discuss and work out mutually beneficial agreements which can lead to substantive normalisation in all fields, particularly in the fields of telecommunications, civil aviation, travel, trade and economic cooperation. For this purpose they reiterated their faith in the method of settling all differences by peaceful means through bilateral discussions.
12. In the course of their talks, the Prime Ministers expressed deep satisfaction at the growing ties of cooperation between their two countries based upon mutual respect, sovereignty, equality and non-interference in each other's internal affairs.
13. The Prime Ministers were glad to note that substantial progress had been made in all spheres of Indo-Bangladesh relations. They reaffirmed their resolve to further consolidate and expand this cooperation, thus fulfilling the deeply felt aspirations of the peoples of the two countries.
14. They expressed the hope that this would promote cooperation in the region as a whole. The two prime ministers reviewed the commercial relations and economic and technical cooperation between the two countries and took the following decisions:

    (a) In recognition of the vital importance of jute in the economies of the two countries and the need for closer cooperation between them in the field of production, trade, technical development, manufacture and promotion of this sector, a Joint Commission at the level of Ministers will be set up.

(b) Noting that, despite measures taken in the past to eliminate contravention of customs and currency regulations of the two countries, unauthorised transactions still continue, a high powered Joint Committee will be set up at Ministers' level to take further immediate measures to effectively control illegal transactions. The committee will ensure coordinated and intensified measures against such transactions and especially against all those engaged in financing and organising such activities.

(c) With regard to the current Balanced Trade and Payments Arrangement, the two Governments will step up exports to ensure the fulfilment of targets.

(d) As a further step towards greater economic cooperation between the two counties, the two governments will establish the following four industrial projects based on the supply of raw materials and products from one country to the other and a guaranteed off-take of the exportable surplus of these projects on mutually acceptable terms and conditions:

   (i) A Cement Plant at Chatak in Bangladesh based on limestone from Meghalaya in India.
   (ii) A Clinker Plant in Meghalaya for supplying Clinker to Bangladesh.
   (iii) A Fertiliser Plant in Bangladesh for the supply of urea to India.
   (iv) A Sponge Iron Plant in Bangladesh based on the supply of iron ore from India.

India will extend credits to Bangladesh on mutually acceptable terms and conditions to finance the procurement of equipment, goods and services as may be available from India for these projects. The modalities of implementing the aforementioned decisions will be worked out immediately by the officials of the two Governments and a programme of action submitted to their respective Governments.

Other areas of Industrial Cooperation might be examined by the two Governments with a view to identifying further specific projects.

15. The Prime Minister of Bangladesh noted with satisfaction that during the financial year 1974–1975, As. 38 crores of credits would be available from India to Bangladesh. Of this, As. 5 crores would be on a Government-to-Government basis for the New Chatak Cement Plant and a similar credit of Rs. 10 crores for the purchase of commodities by Bangladesh from India. The rest of the credit of Rs 23 crores would be provided by the Industrial Development Bank of India.

16. Agreement was reached between the two Governments to carry out a joint survey for the provision of a rail link from Akhaura/Singarbil to Agartala to facilitate the movement of cross traffic from Calcutta to Agartala via Chandpur/Akhaura and vice versa.

17. The two Prime Ministers took note of the fact that the Farakka Barrage Project would be commissioned before the end of 1974. They recognised that during the periods of minimum flow in the Ganga, there might not be enough water to meet the needs of the Calcutta Port and the full requirements of Bangladesh and

therefore, the fair weather flow of the Ganga in the lean months would have to be augmented to meet the requirements of the two countries. It was agreed that the problems should be approached with understanding so that the interests of both the countries are reconciled and the difficulties removed in a spirit of friendship and cooperation. It was, accordingly, decided that the best means of such augmentation through optimum utilisation of the water resources of the region available to the two countries should be studied by the Joint Rivers Commission. The Commission should make suitable recommendations to meet the requirements of both the countries.

18. It was recognised that it would take some years to give effect to the recommendations of the Commission as accepted by the two Governments. In the meantime, the two sides expressed their determination that before the Farakka Project is commissioned they would arrive at a mutually acceptable allocation of the water available during the periods of minimum flow in the Ganga.
19. The discussions of the two Prime Ministers have led to an agreement on the demarcation of the remaining portion of the Indo-Bangladesh border. They expressed their gratification that they were able, thus, to resolve issues that had eluded solution for a whole generation.
20. Reviewing the international situation, the two Prime Ministers reaffirmed that the policy of Non-alignment to which both countries were deeply committed constituted a positive force in the strengthening of national sovereignty, independence and the attainment of peace, stability and a just international economic order, in accordance with the aspirations of peoples the world over. In this context they reiterated their determination to cooperate with each other in making the maximum contribution to the implementation of the decisions of the Fourth Summit Conference of the Non-aligned Nations held in Algiers.
21. The Prime Ministers expressed high appreciation of the work of the Bureau of the Non-aligned countries in its recent conference as well as of the Special Session of the General Assembly convened at the instance of President Boumedienne of Algeria.
22. The two Prime Ministers expressed complete agreement on the desirability of ensuring the sovereignty of countries over their natural resources and an equitable return to the producers of raw material and primary products. They welcomed the Special Programme adopted at the recently concluded Special Session of the UN General Assembly to give unmediated and long-term relief to the developing countries most seriously affected and expressed the hope that this Programme would be implemented expeditiously.
23. The two Prime Ministers reaffirmed their conviction that a lasting settlement of the West Asia crisis could only be achieved on the basis of complete withdrawal of Israeli forces from all occupied Arab territories and the restoration of the legitimate rights of the people of Palestine.
24. They also reaffirmed their support to the right of the Vietnamese people to decide their own future without outside interference.

25. The two Prime Ministers expressed their support for the peoples of Asia, Africa and Latin America in their struggle against colonialism, racial discrimination and the policy of apartheid.
26. Concerning recent developments in the extension of military and naval facilities in the Island to Diego Garcia, the two Prime Ministers considered that the aggravation of big power rivalries in the Indian Ocean runs counter to the concept of the area as a zone of peace and disregards the just aspirations of the littoral States. They expressed the hope that the big powers would respect the resolutions of the UN General Assembly on the subject and the wishes of the littoral States.
27. The Bangladesh Prime Minister expressed his warm thanks for the cordial and friendly reception accorded to him and his delegation and for the demonstration of abiding friendship by the people of India for the people of Bangladesh. He thanked the Prime Minister of India for the warm hospitality extended to the Bangladesh delegation and for the excellence of the arrangements for the visit.
28. The two sides expressed their confidence that the visit of Bangabandhu Sheikh Mujibur Rahman had made a significant contribution to the growth of friendship and fruitful cooperation between India and Bangladesh and to the cause of durable peace in the sub-continent.
29. Bangabandhu Sheikh Mujibur Rahman gave expression to the pleasure with which the Government and people of Bangladesh looked forward to the forthcoming visit of His Excellency Shri V.V. Giri to Bangladesh.
30. Bangabandhu Sheikh Mujibur Rahman conveyed an invitation to Shrimati Indira Gandhi, Prime Minister of India, to visit Bangladesh which was accepted with pleasure.

References

Abbas BM (1984) The Pakistan phase (1951–1971). In: The ganges water dispute. The University Press, Dhaka, pp 27–28

Ahmed R (2001) The emergence of Bengal Muslims. In: Ahmed R (ed) Understanding the Bengal Muslims: interpretive essays. UPL, Dhaka

Avtar SB (1996) India-Bangladesh relations 1971–1994. In: Documents, vol 1, Siba Exim, Delhi, pp 91–92

Ayoob M, Krishnaswamy S (1972) The liberation war. S Chnad, New Delhi

Bhasin AS (1996) India Bangladesh relations documents 1971–1994, vol 1, Siba Exim, Delhi

Chakravarthtty N (1986) Bangladesh. In: Bajpai US (ed) India and its neighbourhood. Lancer International, New Delhi, pp 274–278

Crow B et al (1995) Sharing the ganges the politics and technology of river development. Sage, New Delhi

Dhiren M (1972) Indira speaks on genocide war and Bangladesh. Academic publishers, Calcutta, p 94

Harun Ur Rasid (2002) Bangladesh relations with India (after 1975). In: Indo-Bangladesh relations an insider's view. Har-Anand Publications, New Delhi, p 62

Hasan S (1991) Environmental concerns in South Asia. In: The international institute for strategic studies. Adelphi Papers, London, p 35
Hasan Z (1994) The separation of East Pakistan: the rise and realization of Bengali Muslim nationalism. OUP, Karachi
Iftekharuzzaman (1994) The ganges water sharing issue. BIISS J 15(3):221–222
Indira Gandhi (1980–1981) In: Selected speeches and writings, vol 4, p 552
Indira Gandhi (1984) In: Selected speeches and writings, vol 3 (1972–1977). Publications division, New Delhi, p 72
Islam N (2003) Making of a nation: Bangladesh. University Press Limited, Dhaka
Iyer R (2003) Conflict resolution: three river water treaties. In: Water perspectives, issues, concerns. Sage, New Delhi, pp 236–238
JRC Talks Fail (1980) Meetings at other levels if necessary. Bangladesh times, 12 July
Khan S (1975) In: East Pakistan to Bangladesh. South Asia Books, Columbia, MO
Khosla IP (2005) Bangladesh–India relations. S Asia J 4(3):73
Maniruzzaman T (1980) General Zia's regime. In: The Bangladesh revolution and its aftermath. Bangladesh Books International, Dacca, p 209
Mehta J (2010) Catapulted to foreign secretary. In: Tryst betrayed: reflections on diplomacy and development. Penguin Books India, Delhi, p 221
Muni SD (1980) South Asia. In: Ayoob M (ed) Conflict and intervention in the third world, Palgrave, London
Rangachari R, Iyer RR (1994) Indo–Bangladesh Talks on "The Ganges Waters Issue". In: Verghese BG, Iyer RR (eds) Harnessing the Eastern Himalayan rivers regional cooperation in South Asia, Konark Publishers, New Delhi, pp 173–174
Sobhan F (ed) (2002) Bangladesh India relations: perspectives from civil society dialogue. Dhaka University Press, Dhaka, p 63
Surjit M (1984) Cooperation, strife or hegemony? India and its smaller neighbours, India's search for power Indira Gandhi's foreign policy 1966–1982. Sage, New Delhi, p 277
Swain A (2012) Global climate change and challenges for international river agreements. Int J Sustain Soc 4(1/2):72–87
Tariq K (1998) The Bangladesh-India treaty on sharing of the Ganges waters: genesis and significance. BIISS J 19(2):222
Verghese BG (1999) Waters of hope from vision to reality in Himalaya–Ganges development cooperation. Oxford University Press, New Delhi, pp 391–392
Verghese BG (2001) From dispute to dialogue to doing. In: Biswas AK, Uitto Juha I (eds) Sustainable development of the Ganges Brhmaputra-Meghana-Basins. United Nations University Press, Tokyo, p 173
Verghese BG, Iyer RR (eds) (1993) Harnessing the Eastern Himalayan rivers regional cooperation in South Asia. Centre for Policy Research with Konark Publishers, New Delhi
Wilcox WA (1973) The emergence of Bangladesh: problems and opportunities for a redefined American policy in South Asia, American Enterprise Institute, Washington

Chapter 2
Domestic Variables: Compulsions of Competitive Politics

The Ganges dispute embodies not only templates of hydropolitics but also represents historical, political, cultural and social perceptions and prejudices of the two nation-states. These aspects have shaped the relationship of two neighbours to what it is today—warm and cold at same time. They are characterised as domestic variables. The discussion on these aspects is warranted because they are underpinnings on which lay the government's negotiation process and its outcomes. By quirk of fate, India is present in the domestic political discourse of Bangladesh which has played an important role in the bilateral relationship in general and Ganges river water dispute in particular.

India and Bangladesh have had a long-shared social, cultural and political history. Bengalis of east and west lived together for ages till partitioned into two entities in 1905 and 1947. The history of relationship between the two communities —Hindus and Muslims—having separate belief system, mostly belonging to unequal economic strata and numerical disparity led to have different political aspirations in both wings. The simmering tensions between the two communities culminated into partition of British India. Muslim majority east Bengal became part of newly formed Pakistan.

The present chapter is structured in the following manner. The first section discusses the common history being shared by India and Bangladesh. The second section deals with emergence of India as a factor in Bangladesh politics. The following section examines evolution of Indian federalism and its impact on Indo-Bangladesh relationship. The last section highlights deliberations of the Parliament on the Ganges. The chapter ends with the conclusion.

P. Pandey, *India Bangladesh Domestic Politics*,
DOI 10.1007/978-981-10-2371-2_2

2.1 Shared History and Divergent Aspirations

Here history is employed in an explanatory role to understand India and Bangladesh relationship. Hindus and Muslims lived together for ages but incidents from 1757 to 1947 were critical in deepening and widening the gulf between them.[1] Though it is difficult to cover a long history in such a small space, only important events have been selected to demonstrate the gradual consolidation of two communities along religious lines. The British East India Company made Calcutta centre of business activities which overshadowed other parts of Bengal. While Calcutta had 35 million population, Dacca (the present day Dhaka), the next largest city after Calcutta, had only about 51,000 inhabitants. Demographically, there was almost equal numeric balance between these two major religious communities in both parts, more Hindu rural population to the West and more Muslim population to the east.

Following the Battle of Plassey in 1757, the Permanent Settlement was introduced which granted the landlord the right to collect land revenue in Bengal. This enhanced the dominance of Calcutta. Under the new settlement, *Zamindars* (landlords) became virtual owner of the land. It was the Zamindar who collected taxes from the tillers. The rate of rent was fixed. The zamindars were bound to pay their taxes without failure. The Permanent Settlement act engendered communal feelings because most of the zamindars were Hindus and most of the ryots (tillers) were Muslims in East Bengal. This consolidated the two communities on communal lines in spite of the fact that the issues involved were about economics of rent.

Another major incident was the *first war of independence* or *sepoy mutiny* in 1857.[2] Following this, the administration was taken over by the British government from East India Company in 1858. The British announced major changes in the Indian administration. Lord Ripon brought a very limited elective principle through his act on local government. The franchise rules mandated either property ownership or educational attainments necessary for participation in voting. This was understood by the Muslims as unfavourable to them because of the stringent franchise rules. Even Muslim dominated areas were not going to be benefitted by this. Another important consequence was that the British divided Indian population into martial and non-martial races which further deepened the distrust between Hindus and Muslims.[3] Lord Curzon partitioned Bengal in the name of administrative convenience. According to this new arrangement, whereas the eastern

---

[1]Good description of about cultural transformation of the period is given in Joya (1994), Harun ar Rashid (2003), Iqbal (2010).

[2]There is a disagreement among the historians whether to call this as first war of Indian independence or sepoy mutiny.

[3]Martial races means those who were considered brave and well-built for fighting. Non-martial were those according to the British unfit for fighting.

province was joined with Assam into the province of eastern Bengal and Assam; the remaining area became the province of West Bengal, Bihar and Orissa, each with a Lieutenant Governor.

This decision was fiercely opposed by the Hindu Bengalis. Some of the important Hindu Bengalis like poet Rabindranath Tagore and Aurobindo Ghosh lent their voices and pens to the protest. Lord Curzon toured eastern side of Bengal to convince and explain Muslim leaders how they would benefit from the decision. The Muslims of east Bengal took the partition very favourably. In 1906, a group of leading Muslims met in Dhaka at the invitation of Nawab Salimullah of Dhaka where they laid the foundation of the All-India Muslim League. The basic goals of the Muslim League were to support the Crown and to work for the interests of the community in India. The League saw a great opportunity in the partition of Bengal. Soon, the eastern Bengal Muslims began realising the advantages of being a separate province.

Muslims supported the partition because they saw in it the opportunity to make Dhaka as the centre of activity. Soon, it became the bustling city with full of activities. Muslims particularly the Muslim peasants gave a call to boycott campaign, made anti-zamindar and anti-money lenders' slogans, who were mostly Hindus which, in turn, led to a number of communal disturbances in some of the districts of eastern Bengal and Assam during 1907. The Swadeshi movement was started by the anti-partition group which opposed the British decision to partition Bengal. These leaders of the Swadeshi movement spoke for all Bengalis but the symbols they used and the achievements they referred to were predominantly of Hindus. The words used by Bankim Chandra Chatterjee, Vivekananda, Keshub Sen, Bipin Pal, Surendranath Banerjea or Romesh Dutt did not appeal to Muslims.

In 1911, the British decided to annul the partition of Bengal. For the moderates of Bengal, the nullification of the partition was the cherished goal of many years of petitioning, and they assumed their political methods vindicated. On the other hand, the Muslims of Bengal felt betrayed. Nawab Salimullah expressed his anger over the decision at the Calcutta session of the Muslim League in the presidential address, "to us, the Musalmans of East Bengal, the annulment means the deprivation of those splendid opportunities at self-improvement which we had secured by the Partition".[4]

Another incident which further deepened the gulf between the two communities was the Communal Award of 1932 which accepted most of the demands of the Muslims. While this award was accepted by the Muslim League, Indian National Congress opposed it. Finally on March 2, 1940, in Lahore, the Muslim League passed a resolution which is called the "Pakistan resolution". Fazlul Haq, an important Bengali leader, was among the movers of the resolution. In the 1946 election, Bengali Muslims voted overwhelmingly for the Muslim League. Among

---

[4]Quoted in (Ahmed 2004), p. 278.

Muslims, the League received 82 % of the votes, polling over 2 million Muslim votes and capturing 114 out of 121 Muslim seats, the highest for any of the Muslim-majority provinces.[5] Thus, the 1946 elections ensured a Muslim League government in Bengal. H.S. Suhrawardy became the Prime Minister of Bengal and remained in that position till 1947.

Immediately, after the decision of the British to withdraw from India, partition of territories into two entities became urgent and critical. A Commission was formed by the British to be headed by Sir (later Lord) Cyril Radcliffe. The Muslim League demanded that the Bengal province should be partitioned on the basis of contiguity and majority principle.[6] Thus, it was decided to "demarcate the boundaries of the two parts (the province) on the basis of ascertaining contiguous majority areas of Muslims and non-Muslims while taking into other factors".[7] This principle was followed for division of Bengal except in two cases. The representatives of Hindus insisted that Murshidabad be included in West Bengal although it was a Muslim-majority district. According to continuous majority principle, Murshidabad should have gone to Pakistan and Khulna, a Hindu-majority province, should have come to India. But decision was taken otherwise. It was generally agreed that the survival of Hooghly as a port (and of Calcutta as an entrepot of trade) depended on its link with the river Ganges, which flowed through the northern edge of Murshidabad. The commission accepted the demand of the Congress that Murshidabad should be given to West Bengal.[8] It is generally seen as a bargain for Khulna, a large Hindu-majority district to the east of 24 Parganas, for Murshidabad.[9]

This is the brief historical background of the circumstances leading to the partition of India. Though this is not detailed description of incidents leading to partition, it highlights palpable tensions prevailing in the society which finally led to total rupture in 1947. East Bengal became eastern part of Pakistan on religious grounds. But very soon, relationship between east and west Pakistan soured. After a great loss of lives in fierce military clampdown carried by the Pakistani forces, east Pakistan emerged as Bangladesh, an independent sovereign entity in 1971. India played a critical role in the struggle of Bengalis to become an independent nation. But the Indian participation has not been appreciated unequivocally.

Since India and Bangladesh are closely connected, domestic changes in India make a big impact on neighbour. Indian federation, which will be explained below, has evolved from a mute spectator to active shaper of foreign policy that had bearing on the construction of the farakka barrage.

---

[5]Harun-or-Rashid, op. cit., pp. 214–215.

[6]Chakraborty (1974), p. 53.

[7]Quoted in (Joya 1999), p. 196.

[8]Chatterji op. cit., 1999, p. 215.

[9]Ibid., p. 210.

2.2 Compulsions of Indian Federalism

2.2.1 West Bengal Being Conscious of Its Position

Immediately after independence, West Bengal was affected by policy changes of the central government and separation of east Bengal. Economically, as in so many other spheres, there was an age-old interdependence between the two Bengalis. With partition, this interdependence was breached. Business activities had developed in West Bengal during direct colonial rule, but their main market was east Bengal (two thirds of old Bengal). Thus, West Bengal has to bear the brunt of the loss of market and huge rush of migration of Bengalis from Pakistan. The central government was criticised for not giving attention to migration issue from the eastern side as much as it did to the western side of the border.[10] East Bengal was the main centre of jute production. After partition, West Bengal's share of the jute duty was cut by the central government on the ground of loss of the jute-growing areas to Pakistan, thus the erosion of revenue share. West Bengal's share of the income tax was also reduced on the ground of the loss of population and area as a result of partition.[11] All these policy changes did not go down well with people of West Bengal. "People of West Bengal widely assumed that the central government has been discriminating against them. Policies in the form of excise or what, financial policies of the centre discriminated against the state and in favour of states like Maharashtra and Tamil Nadu".[12] West Bengal's industrial and commercial interests were seriously affected.

As discussed in the first chapter, many commissions and committees had already concluded that the Farakka barrage should be constructed to solve the problem of the Calcutta port. Port maintenance work comes under the central government according to the provisions of the constitution. Owners of shipping and commercial interests had been protesting against the decreasing draught of the Hooghly river.[13] They felt that their interests were not taken care of by the central government especially when Man Singh Committee (constituted to study the possibility of Farakka barrage) had already recommended the construction of barrage in 1952 only. From 1953 onwards, Members of Parliament began to question the Central government's delay in the construction of the Farakka barrage regularly. West Bengal government also demanded the central government to construct the Farakka barrage to desilt the port of Calcutta.

Members of Parliament (MP) from West Bengal began pressurising the central government for the Farakka barrage regularly. In 1954, in one of the discussions, a West Bengal Member of Parliament made the Deputy Minister of Railways and

---

[10]Ghosh (2011), p. 387–388.

[11]Roy (1971), p. 36.

[12]This argument has been developed by Roy (1971).

[13]Draught is defined as the depth of a ship by her outer line up to which she submerges in water with safety.

Transport to accept that the River Hooghly had been deteriorating as a result of "lack of perennial upland discharge" and the "play of natural tidal forces". In return, the minister conceded that the Ganges Barrage scheme had been under consideration.[14] Again in 1955, another West Bengal MP tried to increase the pressure on the government by asking: is it not a fact that the research station (The Central Water and Power research station) at Poona has commented that unless the Ganges Barrage scheme is taken up, the passage of the Calcutta Port is not safe for transportation, and if so, why is the Ganges Barrage not being taken up at the present?…. The Deputy Minister of Railways and Transport Mr. Alagesan replied:

> … There is no use of raising this alarm that the Calcutta Port will be threatened if something is not done. This alarm has been there for a number of years. But the particular question whether there should be a barrage constructed at a particular point is, I think, under the consideration of the Ministry of Irrigation.[15]

Serious discussions were already going on in the West Bengal assembly for construction of the Farakka barrage. Speaking during the 1959 budget debate in the West Bengal Assembly, Bankim Mukherjee, a member for the Communist Party of India, criticised New Delhi's attitude towards West Bengal. "The Farakka Barrage scheme had not yet been taken up for execution though Calcutta Port was dying. It was the Centre's policy he alleged, to move important government offices from Calcutta to Bombay".[16]

Year after year, aggressiveness of the members of Parliament from West Bengal became fierce. Members used all possible Parliamentary tools available to them to influence the central government. In 1956, a 'cut motion' was introduced which proposed a cut in the grant of the Ministry of Irrigation and Power by a nominal Rupees 100 because of the "delay in preparing the Ganges Barrage Scheme".[17] A similar motion was put in place again in 1956.[18] In September 1958, the Minister of Irrigation and Power made a statement on the Ganges Barrage Project before the Lok Sabha, and in November it was debated by the house for two-and-a-half hours.[19]

Thus, it underlines the continuous pressure exerted by members of Parliament from West Bengal in support of demand for the Farakka barrage since 1950s. Another member of Parliament for Berhampore, West Bengal, T.K. Chaudhuri expressed exasperation over delay in the beginning of construction of the Farakka barrage as the 6 years of "surprising procrastination, evasiveness, and

---

[14]India, Lok Sabha Secretariat, A Selection from Questions and Answers in the Lok Sabha (1952–57), Ministry of Transport, New Delhi. March 1957, question on 29/11/54, p. 9.

[15]Lok Sabha Debates (Part-1 Questions and Answers), vol. IV 1955 (25th July to 20th August 1955) Tenth Session, New Delhi: Lok Sabha Secretariat, New Delhi. Ibid, question on 17/8/55.

[16]"Congress-opposition battle of statistics," *The Statesman*, 18/2/59.

[17]India, Lok Sabha Debates, "Demands for grants-Ministry of Irrigation and Power," 3/4/56, Sl, 3, col. 4185.

[18]Ibid., "Demand for grant-Ministry of Irrigation and Power," 31/7/57. Col. 6180 etc.

[19]India, Lok Sabha, Debates, "Ganges Barrage Project (Discussion)" 19/11/58, S2, 22, pp. 589–644.

indecisiveness" since the project had been mooted in 1952. Two more West Bengal MPs delivered lengthy speeches with similar expressions of frustration. One of them, H.N. Mukherjee of Calcutta Central drew the house's attention to the secrecy surrounding an expert report on the project.[20]

In fact, in March 1959, all the MPs belonging to the Congress party representing constituencies of all north-east India, decided to use their combined influence to get the project sanctioned. The Statesman reported:

> A meeting of Congress MPs called today to discuss the Calcutta Port problem came to the conclusion that top priority should be given to the Farakka barrage to enable big ships to continue to come to Calcutta Port. MPs from eastern Uttar Pradesh, West Bengal, Bihar, Assam, Orissa, Manipur, Tripura, and Andaman and Nicobar Islands were present. Members expressed concern at the increasing salinity of the waters of the Bhagirathi River. It was felt that this problem and also the construction of the Sonu Bund and the Rihand Barrage should be examined in relation to the entire region. It was generally agreed that members should try to impress upon the government to take steps to 'save' Calcutta Port and to make efforts for proper utilisation of water resources for the region.[21]

The Bengal Chamber of Commerce was also very vocal in its demand for the Farakka project because their commercial interests were being badly affected. An analysis of the Calcutta port's share in the foreign trade (excluding treasure) of India from 1870–1871 through 1946–1947 showed the important contribution made by this port to the Indian economy. The Calcutta port handled about 41.6–53.2 % of India's export and about 39.0–47.4 % of its imports (both in terms value). In relative terms, gradually, the share of Calcutta port in total traffic handled by major ports of India declined from about 50 % in 1929–1930 to about 43 % in 1947–1948, 23 % in 1964–1965, 11 % in 1977–1978 and 10 % in 1988–1989.[22] The decline of draught of Hooghly increased the cost of export and import through Calcutta Port. Therefore, it affected profits of commerce and industry heavily. Only smaller ships were able to visit the port because of the reduction in the draught when the economics of scale were demanding bigger ships. The Farakka Project was also supported because of positive externalities like providing a solution to the navigational problems of the lower Hooghly, improving condition of salinity of Calcutta's water supply, provide year-round navigational facilities on the Bhagirathi and a rail and road bridge across the Ganges at a strategically useful point.

> Moreover, quantity of cargoes handled by the Calcutta port was going down and down. In 1951, Calcutta was the biggest port handling 94 lakh tonnes a year against 75 lakh tonnes handled by Bombay, the second biggest port then. Later, other ports have got built up or expanded and modernised and are able to handle (and are handling) part of the cargo which passed through Calcutta previously. Before the two droughts and the war with Pakistan in 1964–1965, Bombay handled 189 lakh tonnes against Calcutta's 109 lakh tonnes. Drought and Indo-Pak war in 1965 disrupted the economy. By 1968–1969, Calcutta was handling only 79 lakh tonnes of cargo while Mormugao was 88 lakh tonnes and Visakhapatnam 80 lakh tonnes.[23]

---

[20]Ben Crow & et al., p. 58.

[21]"MPs concern over port problem," The Statesman, 18 March 1959.

[22]Sau (1990), pp. 1015–26.

[23]Ranjit Roy op. cit., p. 32.

In April 1960, even representatives of foreign shipping companies publicly expressed the view that if the deterioration continued for another 2 years the port of Calcutta would be lost. Apart from intensifying dredging services, the long-term solution, according to them, was laid in the supply of fresh water. Only way to achieve this was the early completion of the Ganges Barrage scheme.

Therefore, commercial interests were adding pressure on politicians, members of Parliament and West Bengal government to influence the Central government to go for construction of the Calcutta port. In fact, the Bengal Chambers representing interests of shippers made strong influence on the government. Even in 1833, it was pressure of the Bengal Chambers of Commerce that directed the government of Bengal to establish the Hooghly Commission. Again, during the 1950s, the same Chamber of Commerce was increasingly troubled by the deterioration of the Hooghly. Its president, Mr. J.D.K. Brown told the 1959 Annual General Meeting of the Chamber:

> A matter which has gravely concerned shipping interests in particular and commerce and industry generally in this part of India throughout this past year has been the state of the River Hooghly. This is not entirely a local problem as it has all-India significance when we consider the portion of the export and import trade which passes through the Port of Calcutta …it appears that unless something is done, and done as quickly as possible, the repercussions on the trade and industry of North East India and of India itself might be very severe.[24]

All these demands were putting great pressure on the government. Apart from members of Parliament, the provincial government of West Bengal lobbied hard with the Central government for construction of the Farakka barrage. This can be illustrated with the citation of exchange of letters the West Bengal Chief Minister, Dr. B.C. Roy had with the Indian Prime Minister Jawaharlal Nehru. Dr. B.C. Roy wrote a letter to the Indian Prime Minister Jawaharlal Nehru in March 1952 in which he underlined the need for construction of a barrage at the top of Murshidabad district so as to divert the water from other branches of the Ganges into the Bhagirathi and then into the Hooghly. "Calcutta port must be saved from being silted up by the saline water from the Bay of Bengal. Further, the banks of the river Hooghly are studded with a very large number of mills whose contribution to the wealth of the nation is not negligible". Further referring to the bridge-cum-barrage project at Farakka which was essential for the establishment of communications between the lower and southern parts of West Bengal, in the same letter he underlined: "Barrage at Farakka would keep the Hooghly alive, a factor on which depended the future trade and prosperity of the State".[25]

The Chief Minister had been urging for the Ganges Barrage Project since 1954–1955 when the Second Five Year Plan was about to begin. Three letters were

[24]Quoted in India (1961), p. 60.

[25]Roy, op. cit., pp. 205–207.

exchanged between the Prime Minister and the Chief Minister. The central government was aware of the situation and intended to intensify the work at Farakka as early as possible. In his letter of the March 12, the Prime Minister gave a categorical assurance that Ganges Barrage scheme was "certainly being included in our plans".[26]

Further, in a letter dated March 8, 1960, the Chief Minister wrote to the Prime Minister Nehru about the benefit of having a barrage at Farakka:

> I can only tell you that this scheme is essential not merely for the economy of the state itself but also for the safety of the Port of Calcutta which handles a very large quantity of goods for export and import purposes. It will save the city of Calcutta from extra salinity during the hot weather. Meanwhile it will help us to have balanced distribution of water in the Delta and to the different parts of West Bengal.[27]

As discussed earlier, a number of committees were set up by the British as well as the Indian government after independence most of them were coming to the same conclusion of the construction of a barrage for improving the Calcutta port. By the end of 1960s, a consensus had emerged among the politicians, wider public and the businessmen about the urgency of the Farakka barrage. Expressing his frustration about setting up of number of committees to study the barrage possibility and inability to come with a clear cut decision resulting into further delay, the Chief Minister wrote,

> Commissions have come and commissions have gone: enquiries have been made several times, perhaps 'ad nanseam'. Now I understand that the Planning Commission is not satisfied with the results of various enquiries that have been made in the past and that they want to have another enquiry by another foreign expert. Meanwhile, what is happening is that the East Pakistan Government have already taken 8,000 cusecs of water from the Ganges for their Kobodak Scheme and they will very soon increase the amount of their intake by pumping water from the Ganges to Kobodak up to 20,000 cusecs. This means that we shall be lagging far behind… I can only tell you that this Scheme is essential not merely for the economy of the State itself but also for the safety of the Port of Calcutta which handles a very large quantity of goods for export and import purposes. It will save the city of Calcutta from extra salinity during the hot weather. Meanwhile, it will help us to have a balanced distribution of water in the Delta and to the different parts of West Bengal. I urge upon you to take this matter up seriously.[28]

Amidst clamour for the Farakka barrage in India, Pakistan was demanding a ministerial level dialogue between India and Pakistan, Pt. Nehru replied

> …it is always desirable, where interests conflict, like river interests or water interests, to discuss them and try to find a way which is satisfactory to both. Now, if this done at the officers level, the officers' naturally, are tied down to certain, shall I say fixed policies as given down in the brief; there is no room to vary them. While ministers can find out these ways which are advantageous to both. But before all this is done, the facts must be known.

---

[26]Saroj Chakraborty, op. cit., 444–445.

[27]Nehru-Roy correspondence in Chakraborty, op. cit., pp. 444–47.

[28]Ibid., pp. 446–447.

> These are exceedingly complicated things. It is not a simple affair. When you go to any kind of a scheme of the river water or canal waters, it is frightfully complicated.[29]

Thus, the decision to construct the Farakka barrage was taken after lots of investigation and political debates. Thus, West Bengal was really concerned about losing its shipping interests, most importantly navigation.[30] Since West Bengal already lost pride of place to Maharashtra and other Southern provinces in the industry and commerce; it was really particular to maintain its interests at any cost, so it became politically important for state government as well as political parties of the state to get the Farakka barrage completed as soon as possible. K.K. Framji, who had played an important role in the preparation for the project, referred to the Farakka completion as the "fulfilment of a dream". He writes that "the Public Accounts Committee of the Lok Sabha wrote in 1976 that the dedication of the project was "for many in the country … almost like a dream come true. Hopes long deferred now seemed near fulfilment".[31] As mentioned in the first chapter, Pakistan was attributing an evil design behind the construction of the Farakka barrage. The whole discussion underlines that it became impossible for the central government to delay the project any more. Amidst domestic demand for the Farakka construction, Pakistan's internal problem began to spill over India.

People from East Pakistan began fleeing to India to avoid atrocities committed by the Pakistan army. There were almost seven million refugees from East Pakistan to Assam, Tripura and West Bengal causing massive demographic pressures on these states. Huge Indian resources were diverted for giving shelter to these people.[32] Keeping large number of refugees in mind, a separate department was set up to deal with East Pakistan refugees.[33] Gradually, a strenuous pressure was being felt on exchequer because of influx of large number of population. Mrs. Indira Gandhi, the Prime Minister of India made a statement to Parliament "On present estimates, the cost to the Central Exchequer on relief alone may exceed Rs. 180 crores for a period of 6 months. All this, as Hon. Members will appreciate, has imposed an unexpected burden on us".[34] Indian involvement in Bangladesh's independence positioned India in the domestic politics of the country post-independence period which will be discussed below.

---

[29]Pt. Nehru replying on Farakka Barrage regarding whether ministerial meeting will affect Farakka barrage or not, Lok Sabha Debates Second Series, vol. LVI, August 7–19, 1961, Fourteenth Session, New Delhi, Lok Sabha Secretariat.

[30]One of the major critics of this proposal was Kapil Bhattarya, an eminent engineer. He was categorically affirmed that it would be impossible to get 40,000 cusecs water during lean season. Thus the barrage will definitely fail in its promise of navigability.

[31]Quoted in Ben Crow and et al., op. cit., p. 63.

[32]All these provinces of India share border with Bangladesh.

[33]Dixit, (1999), p. 49.

[34]Indira Gandhi (1972), p. 16.

2.3 India as a Factor in Bangladesh Politics

The Bangla Government in exile was formed and located in West Bengal in India. With the surrender of Pakistan army to Indian forces, Bangladesh became an independent nation-state in 1971.[35] With this, relationship between India and Bangladesh embarked on a new fervour and flavour. At first, the association was full of enthusiasm and hopes from each other. Mujib considered India to be the closest ally of Bangladesh. Provision of the Friendship Treaty confirmed Indo-Bangladesh special relationship. Article 8 of the treaty put a condition that Bangladesh should not enter into or participate in any military alliance directed against India. Another article, Article 9 committed the two parties to refrain from giving any assistance to any third party in armed conflict. In her speech at a public rally in Dhaka on March 17, 1972, Mrs. Gandhi tried to clear India's intentions to help Bangladesh liberation:

> If India has helped you, it is because we could not sit idle after hearing your voice and after knowing of the sorrow and suffering that you have undergone. If we have helped you, it is in order to be true to ourselves, and to the principles to which we have adhered for years. I trust that, in the coming years, friendship between our two countries will be built not on the basis of the assistance that might have given to you now, but on the basis of the full equality and mutual benefit of two free and sovereign nations….If we offer you cooperation, it is not out of any desire to weild influence over you. We want you to stand on your own legs.[36]

But this did not diminish the misgivings which had grown against India, among different groups of people which would be discussed shortly. Like Indian, Bangladesh became a secular state and all political parties with Islamic leanings were banned from the day the Awami League assumed the office in Dhaka. This ban did not resonate with people of Bangladesh and this was widely seen as an Indian ploy to retain its influence in Bangladesh. "Bengali Muslims had not been serious in stressing a Muslim identity in Pakistan politics, but they seemed more anxious to do so in secular Bangladesh!" Islamic identity became the rallying point of resistance against Indian predominance in Bangladesh politics. In the summer of 1973, the so-called Muslim Bengal movement was started by the right-wing Islamic forces to revive Muslim sentiment and to resist the secularism propounded by the ruling Awami League.[37]

Maulana Bhashani, leader of a five-party United Front, opposed Mujib's Awami League government. Within a few months of Bangladesh's independence, the Maulana began open and bitter criticism of Indian influence in Bangladesh. He also criticised the Awami League's policy of secularism and gave tacit support to the "Muslim Bengal" movement by saying that within 6 months' time there would be a

---

[35]For details on Indo-Pak war of 1971 and violence of 9 months by the Pakistan army see.

[36]Speech of the Prime Minister of India, Mrs Indira Gandhi, at a public rally in Dhaka, 17 arch, 1972, Bhasin.

[37]Rashiduzzaman (1977), p. 797.

"new flag" in Bangladesh, a euphemistic reference to the flag of Muslim Bengal. But this movement subsided gradually when several of its prominent leaders were arrested for clandestine activities. In December 1972, Bhashani called upon the Bangla government to repeal all pacts and treaties that Pakistan had signed with India before the 1970 election. In January 1973, he was further emboldened and claimed that some organisations in India were conspiring to integrate Bangladesh with India and demanded the resignation of the Awami League government and the formation of a National government. In July 1973, he was threatening to launch a jihad against India in the form of a boycott of Indian goods. Later when India detonated nuclear test explosion he characterised it as a "veiled threat to her smaller neighbours", and a deviation from the path of non-violence.[38]

Thus, Maulana Bhashani became the central figure of attraction for people who were against India and the Awami League government. The pro-Islamic forces began to rally around the octogerian leader Maulana Bhashani, who professed socialism but was believer of Islamic principles. He began advocating Islamic socialism. As the law and order situation deteriorated after a devastating flood followed by widespread famine, the Awami League leaders began to blame the so-called pro-Pakistani elements and radicals for the secret killings and political violence in Bangladesh.

India was criticised that it did not withdraw troops from Bangladesh immediately after its liberation. But the discussion in Indian Parliament and general Indian government's opinion were in favour of withdrawing troops from Bangladesh at the earliest. On being questioned that some countries are demanding that India should keep the armed forces in Bangladesh for some time, Mrs. Gandhi said in reply that

> …it is right for the Indian army to move out as soon as it possibly can. Already quite a large number have come out and the others, I hope, will come out soon. As I said earlier, there was some fear about the non-Bengalis who are resident there, about their safety. But I think the Bangladesh Government is in full control of the situation.[39]

On the issue of the cost of Indian army who were still stationed in Bangladesh, the Defence Minister replied that

> in rcsponse to a request from the Bangladesh Government, a small detachment of the Indian Army deputed to assist the Bangladesh forces in carrying out anti-insurgency operations in the Chittagong Hill Tracts during the period March 15 to May 15,1972. This detachment was withdrawn from Bangladesh by May 15, 1972. Since then no Defence Service personnel other than those employed in our High Commission in Dhaka are in Bangladesh. The cost has been borne by the Government of India.[40]

---

[38]Kodikara (1993), p. 134.

[39]Speaking in the Press Conference in New Delhi, December 31, 1971 Retrospect and Prospect, in Indira Gandhi India and Bangladesh Selected Speeches and Statements March to December 1971, New Delhi: Orient Longman, 1972, p. 175.

[40]Bhasin (1996), p. 46.

Whether India was interested or not to influence Bangladesh government and society, Indian shadow was writ large on all aspects of the emergent country because of long association.

The numerous government officials, teachers, students, politicians, and prominent individuals who had spent months in India during the 1971 liberation struggle were inclined to accept Indians in Bangladesh for various activities. In fact, they did not hesitate to raise the factor of Pakistani collaboration against those who did not agree with them. For a while, to some, patriotism was sarcastically linked with "pilgrimage" to India. The public reaction was quick. One Dhaka newspaper wrote in disgust that Bangladesh was a nation of 65 million collaborators (those who collaborated with Pakistan during liberation movement). The returnees from India were sarcastically called "Hajis" by the critics of the Awami League government. The Indian presence was also conspicuous by the visits of various Indian high officials and distinguished private individuals to help the nascent state and cement friendship between the two countries.[41]

Bengalis were not in agreement about the nature of state that Bangladesh should adopt. The reason behind acrimonious relationship among these groups was that Bangladesh society was divided among various factions depending on their role in the liberation struggle: those who had fought the liberation war and, those who were confined to Dhaka and, the collaborators of the Pakistan army. Groups like Kader Bahini had independently fought the liberation war and its leaders such as Kader Siddiqui and Abu Taher had waged an armed struggle in areas like Tangail. These groups suspected that the Indian objective to help Bangladesh was to divide Pakistan and gain strategic and economic advantages from the emergence of Bangladesh. The presence of the Indian Army during the surrender ceremony of the defeated Pakistan army did not go down well with them especially when no senior officer of Bangladeshi forces was present.[42] With these kinds of ideological streams prevailing, Indian role in Bangladesh's liberation became contested; consequently, the bilateral relationship was coming under increasing public scrutiny and criticism.

There were widespread allegations in Bangladesh that officials in both countries were involved in smuggling relief supplies, jute, rice and fertiliser, and consumer goods. The persistent rumours was working to destroy whatever goodwill India had cultivated in Bangladesh and to undermine Mujib's popularity. Tension escalated rapidly in early 1974 when several Bangladesh military officers claimed they had been dismissed because they had traced smuggling activities to Mujib's family and friends.[43]

---

[41]Rashiduzzaman, op. cit., p. 797.

[42]Since war was fought between two sovereign nation-states, naturally Pakistan army would have surrendered to Indian army but somehow Indian army would have made even symbolically gesture of making Bangladeshi army feel important. In this see, J.N. Dixit, "War and the Birth of Bangladesh," in Liberation and Beyond, op. cit, p. 109.

[43]Franda, op. cit., p. 128.

Gradually, Mujib began to realise the importance of developing friendly relationship with Muslim countries. The Collaborators Act of 1972 was repealed, which gave some legitimacy to those Muslim leaders who were alleged to have been supportive of Pakistan in 1971. Mujib also accepted an invitation to attend the Islamic Conference which was held in Pakistan in 1974.[44]

2.4 Mixing up Internal and External Relationship

2.4.1 India Synonymous of Mujib Government

India increasingly became synonymous of the Mujib government. Whatever omissions were committed by the Mujib regime, India was to be blamed as well.[45] So, India became an equal partner of all the wrongs in every stream of the Bangladesh government very soon.

2.4.1.1 Administrative Divisions

Bureaucracy in Bangladesh was divided between those officers who fled to India during the liberation struggle and were working closely with the provisional Government of Bangladesh and those who were in Pakistan. The Mujib government gave 2 years seniority in promotion to those officers who were closer to the exiled government in 1971, ignoring a principle of seniority. Many senior Bengali officers who stayed back or were detained in Pakistan until 1974 did not get high positions when they returned to Bangladesh. This raised frustrations in the upper sections of the bureaucracy and felt humiliated when they had to report to junior officers.[46] The discrimination against those who did not participate in the liberation struggle was restricted not only to civil services but also to other branches of government.

2.4.1.2 Military Division

Out of the Bengali soldiers who were in the Pakistan army, almost 15,000 of them defected to the freedom movement in 1971 which constituted the core of the Bangladesh liberation army. After liberation, the Bangladesh army was created with the help of those personnel and the Bengali officers who returned from Pakistan. Like the civil service, the military in Bangladesh was divided between the 'patriots' and the alleged collaborators. Many people who fought in the Mukti Bahini were

[44]Rashiduzzaman, op. cit., p. 798.

[45]Hassan (1989), pp. 44–61.

[46]Rashiduzzaman, pp. 794–795.

absorbed in the army by Mujib as an inducement to surrender arms. They were disliked by soldiers who were already let down with restricted pay and reduction of other benefits in the new nation. Another factor which contributed to simmering tensions between the army and the civilian government was that the former seemed to believe that proper recognition has not been given to Army officers who fought the Pakistani forces from Bangladesh. In March 1971, when the Awami League had collapsed under the Pakistan army's crackdown, the formal declaration for complete independence was made by General Ziaur Rahman on behalf of Mujib from Chittagong Radio station. But Zia was not respected in a manner which deserved to a person who played a crucial role in the liberation struggle. Later, this kind of issue took an ugly turn in the competitive politics of Awami League and Bangladesh Nationalist Party.

Another issue was arms which many groups were keeping with them. Several thousand young freedom fighters, not belonging to trained cadres, possessed arms. This presented a challenge to the government. While Mujib appealed to these freedom fighters to surrender arms to government authorities in different parts of the country; public ceremonies were held for the surrender of arms by the Mukti Bahini, most of the guns went underground and most probably into the hands of the anti-social elements. These elements devastated the countryside with violent crimes.[47] Some suspected that arms from Bangladesh had gone into India because Mujibur Rahman's government failed to retrieve all the weapons and ammunition accumulated by private armies, political parties, and individuals during the liberation war of 1971. It is assumed that Mujib distrusted the army because most of the officers in Bangladesh army were trained in Pakistan and were associated with a military regime which was instrumental in suppression of the liberation struggle. The military is suspected to have been and remains 'anti-India' due to long-standing hard feelings over the last stages of the war, when "the Mukti Bahini alleged that their prowess was snatched when the Pakistan forces surrendered to Lt. Gen. Jagjit Singh Aurora of the Indian army, the Mukti Bahini was not represented during the surrender ceremony, "independent Bangladesh inherited not only a politicised military but also an uneasy civil-military equation".[48]

Mujib built his own Rakkhi Bahini (Security Force), trained by the Indian military. The military also resented the creation of the Jatiyo Rakkhi Bahini (JRB), a paramilitary force, at a time when the military itself was being neglected. In January 1974, a law was passed giving the JRB special powers and legal protection. Thus, the military regarded the JRB as a parallel institution and a threat to its corporate interests. Mujib also created a number of armed political groups under the tutelage of, and loyal to, his family members and close associates. The principal groups were the Jubbo (Youth) League led by Sheikh Fazlul Huque Moni (Mujib's nephew); a number of private armies raised by heroes of the liberation war, the most prominent being the collection of Mukti Bahini (Freedom Fighters) led by Kader

---

[47]Ibid., p. 794.

[48]Mohsin (2001), Franda (1975).

Siddiqui, the so-called "Tiger of Tangail". Although in theory Jubbo League and private armies were affiliated to Mujib's political party (the Bangladesh Krishak-Sramic Awami League, or BAKSHAL), most observers likened the bands of armed youth as Mujib's private armed forces.

Opposition parties, of course, attempted to maintain their own armies. The result was a state of almost constant warfare in Bangladesh countryside, with opposition leaders and Awami League members being killed almost every day. Mujib estimated that in 1974 alone more than 3,000 of his supporters, including six members of Parliament, had been killed by 'extremists'. Opposition parties vowed revenge when many of their top leaders were assassinated, allegedly by 'Mujibadis' (believers in 'Mujibism'). Understandably, opposition leaders engaging in this form of guerrilla warfare frequently used India as a refuge. Moreover, they sought to establish contacts with those radical political parties in West Bengal that have for years been convinced of the need for using violence against the Indian government. Thus, radical opposition leaders got united against the governments of both countries.[49]

Moreover, 9-month-long liberation war had thoroughly politicised the forces that were to form the nucleus of the Bangladesh military. The Mukti Bahini liberation force was composed of Bengali troops who had revolted against Pakistan as well as youths who were recruited later. These youths were highly motivated and politically charged. These troops formed the chain of command of the liberation forces and—as part of the former Pakistan military edifice—carried an anti-India psychosis with them. The Mukti Bahini therefore wanted to minimise the role of India, which had become a key factor in the liberation war; the Awami League leadership had not only taken refuge in India, but the provisional government of Bangladesh operated from the headquarters in West Bengal. The military leaders resented the reliance of the political leadership on India. Further, though the war strategy was prepared by the high command of Mukti Bahini, the Indian army was involved in training and arming the troops.

But conflict was not limited to different segments of the government only; tensions were prevailing among different sections of the Awami League also. There were groups like conservatives and radicals within the Awami League which did not get along and followed different paths. Tensions among these groups could be predated to the liberation war, which became severe during the war. Radicals accused the Awami League leaders in India of trying to prevent their entry into the Mukti Bahini (Liberation Army). These Awami League leaders raised questions

[49]It is also understood that rebels and extremist forces of both India and Bangladesh were coming together and were in collusion. It was being believed by the Indian officials that these forces might be getting trained from Beijing though with no clear cut evidence. Gauging the seriousness of the matter, during Mujib's New Delhi visit in 1974, Mrs. Gandhi and Mujibur Rahman agreed to establish a joint ministerial committee to explore ways to coordinate administrative and security forces on the border.

about Mujib Bahini becoming an important component of the Mukti Bahini. They alleged favouritism towards the Mujib Bahini (e.g. in the matter of weapons supply) which undoubtedly exacerbated antagonisms among freedom fighters. These divergences naturally spilled over into post-war Bangladesh.[50] In independent Bangladesh, a split occurred in the Awami League with the radicals forming the JSD (Jatiya Samajtantri Dal or Nationalist Socialist Party).

With tensions increasing among all sections of society, power gradually concentrated in the hands of the Prime Minister and later into the President Mujib ur Rahman. This resulted into emergence of clout of family members of Mujib. "Lots of middlemen came between senior officers and the Prime Minister. So the senior officers have to wait longer to get to meet the Prime Minister".[51] Thus, strong resentment arose among senior officers against Mujib and his influential family members which undermined the unity of the military elite.

Haphazard Situation in Economy

Immediately after liberation, the cost of living jumped from taka (Bangladesh currency) 208 in January 1972 to taka 297 in October. The Mujib regime's economic policies created multiple middlemen in the system. As licenced permit goods were sold and resold, naturally prices were going higher and higher. Before liberation it was expected that there would be mutually beneficial economic cooperation between India and Bangladesh. But after the liberation, the trade agreement and the border trade pact did not live up to people's expectations. The high rates of smuggling further spoiled the economic relations between the two countries. (Because of the high rate of smuggling, trade through the regular channels was minimal. It was blamed that higher prices in Bangladesh were because of smuggling.) Finally, the Bangladesh army was sent to the border to tackle the smuggling problem, and the border trade pact was cancelled. As 1972 drew to a close, major problems in the economy front remained rising prices—especially food prices—labour unrest, multiplication of middlemen and heavy dependence on foreign aid.[52]

Some permissions were granted to businessmen of West Bengal to establish contacts with the private sector in Bangladesh which was interpreted as parts of India's negative and hidden agenda on Bangladesh. This criticism of India found expression in an undercurrent of propaganda against some members of Mujibur Rahman's Cabinet who had been members of the Mujibnagar Government during the liberation war. Finance Minister Tajuddin Ahmed was a particular target of this criticism.[53]

---

[50]Ray (1986).

[51]Ibid., p. 802.

[52]Rounaq (1972), pp. 208–209.

[53]Dixit, op. cit., p. 176.

The rising prices of both manufactured goods and agricultural commodities in India after the liberation were bound to cause inflation in Bangladesh. To reduce the inflation effect, Mujib banned all private trade between India and Bangladesh. At the same time, however, he allowed his relatives to engage in private trade. As his position became the primary subject of scandal in Bangladesh, people of Bangladesh blamed both Mujib and India for their sufferings. By mid-1975, shortly after the coup, a number of Bengalis who had previously advocated close ties with New Delhi began to argue, in the words of a Dhaka University Professor, that "We would prefer the old argument with Pakistan to this arrangement with India".[54]

Controversy between India and Bangladesh was not limited to food prices only. India was criticised in other economic spheres as well. There was an acute shortage of basic clothing in Bangladesh. India agreed to export large number of *lungis* (a cloth worn around waist) and *Sarees* (worn by Women) to Bangladesh to compensate deficit of these items. But India was attacked for sending the sub-standard material to flood Bangladesh and for killing the "Dhaka *Saree* Manufacturing Industry". Rumour spread about Indian malpractices in other fields too. In 1973, when there was a shortage of fish products, certain vernacular newspapers in Bangladesh indulged in the speculation that fishing authorities in West Bengal mixed some chemicals in lower reaches of the river system which flowed through West Bengal before joining the sea which affected the habitat of fishes.[55]

International oil prices also affected Bangladesh. Economic conditions deteriorated in July and August 1974 when serious floods engulfed the country, followed by a cholera epidemic. By 1974 Bangladesh was in the grip of a full scale famine. During 1974 and 1975, the gross domestic product increased by 2 % while population grew by 3 %. The national economy suffered from low productivity, an excessive money supply, deficit financing and galloping inflation.[56] It is said that over 200,000 people died due to famine as the cost of living index went up between 400 and 500 %.[57]

2.5 Assertiveness of Muslim Identity

2.5.1 Complete Reversal

Thus, the above discussion suggests that knowingly or inadvertently India became embedded in the politics of Bangladesh. It was made responsible either for problems of Bangladesh or criticised being associated with the policies of Mujib

[54]Maniruzzaman 2003, p. 67.

[55]Dixit (1973), p. 177.

[56]Jalal (1995), pp 88.

[57]Ahmed (2004), p. 211.

government. As discussed in the first chapter, Mujib and his close family members were assassinated by the military on August 15, 1975 in army coup. Coup brought complete changes in the principles of governance. Whatever the Mujib government professed for governance was reversed by the new regime to show assertiveness of the new government. Signals emanating from India also made it susceptible in the eyes of the new government. General Ziaur Rahman became military law administrator in November 1975 following coups, and then became a civilian president after winning election in April 1977. He formed a political party known as Bangladesh Nationalist Party (BNP)

Pakistan became the first nation to recognise the new regime and its Radio promptly made an announcement that the new government in Dhaka has changed the name of the country from "The People's Republic of Bangladesh" to the "Islamic Republic of Bangladesh". On the contrary, Indian government was unable to communicate to the new government for 3 days because telephone communications between New Delhi and Dhaka was cut off and the Indian High Commissioner to Bangladesh has come to Delhi for consultations. Even journalists in India were unable to go to Dhaka because air links had been suspended. Largely on the basis of speculation, they therefore tended to emphasise the post-coup possibilities for Indo-Bangladesh confrontation. It should be underlined that such speculations were encouraged by the rigid press censorship rules prevailing in India at the time because of the emergency.

When communications between India and Bangladesh were re-established on August 18, 1975, the new government tried to dispel New Delhi's fears. The Foreign Ministry informed New Delhi that the name of the country had not been changed, and President Mushtaque Ahmed stated in official note to the Indian Deputy High Commissioner in Dhaka (communication to New Delhi by telephone) that Bangladesh would "honour all bilateral agreements and obligations", including a 25-year Treaty of Friendship signed by Mujib. Then Mrs. Gandhi called on the Bangladesh High Commissioner in New Delhi and expressed her wish for "success and prosperity to the people of Bangladesh". Twelve days after the coup, India became the 39th nation to recognise the new regime. India was criticised for delaying recognition of the new government for several days after most other major nations—including the United States, Japan, Iran and the Soviet Union—had done so. In India, the Communist Party of India (CPI) proclaimed that Bangladesh coup was engineered by CIA.[58]

By the proclamation of Order I of 1977, the phrase "*Bismillahir-Rahmanir-Rahim*" (in the name of Allah, the Beneficent, the Merciful") was inserted at the beginning of the constitution before the Preamble. With the same proclamation, secularism as one of the state's principle was replaced by assertion of absolute faith in the Almighty Allah. Article 12 which suggests banning communal political

---

[58]Maniruzzaman, op. cit., p. 68.

parties in Bangladesh was also dropped. Article 9, which stressed the linguistic and cultural unity of Bengali nationalism, was likewise omitted. In place of 'Bengali', according new changes, citizens were now to be known as 'Bangladeshis'. These changes were given effect through the Fifth Amendment to the constitution on April 5, 1977. These measures helped Zia to consolidate his power base among the general public and the military. The division between Bengali and Bangladeshi nationalism, however, sowed seeds of discord within the Bangladesh polity. The emphasis of Bengali nationalism which was the basis of the Bangladesh liberation movement was on secularism, whereas Bangladeshi nationalism, though territorial and linguistic, had an overtly religious bias. Not only was the Bengali community divided along nationalist lines, but it also distinguished between the dominant Bengali Muslim community and other religious minorities. Thus, communal politics began to consolidate in Bangladesh.[59]

When Parliamentary election was held in 1979, prominence was given to those men within the new BNP government who had previously been charged with, or convicted of, collaboration with Pakistan during the liberation war. In the 1979 election, more than 400 of the early 2,000 Parliamentary candidates were accused of or convicted of collaboration by Mujib's Awami League regime. Even those who were part of Zia government including three Ministers and four BNP candidates had previously been Ministers in Ayub or Yahya Khan's governments of the 1960s. More than three-fourths of the BNP's victorious MPs in 1979 were said by the Awami League leaders as collaborators including Shah Azizur Rahman, the man Zia appointed as his first Prime Minister after the death of Mashiur Rahman.[60]

After coup, Hussain Muhammad Ershad occupied the power in 1982, he further expanded Zia's policies. Unlike Zia, he did not face any opposition from the military. By 1982, when Ershad took over power in a bloodless coup from the short-lived civilian government of Abdus Sattar, the military had been cleansed of its internal factionalism and had begun to claim a constitutional role for itself. Ershad formalised the entry of military officers into the foreign and civil service on the basis of a quota system. In the Foreign Service, 25 % of the posts were allocated to the armed services. The police forces, too, were brought under the military's influence. Indeed, Ershad literally inhabited the administration with military personnel including important bodies such as the National Economic Council, the Committee of Food, Agriculture, and Rural Development, the Energy and Mineral Resources Committee.

Ershad's policies to India was very similar to those of Zia's, except for a brief period which has been discussed in the previous chapter. But with the growing pressure internally from different political parties, Ershad used India card to divert public attention. These domestic undercurrents overlaid the negotiation strategies of

[59]Franda, op. cit., p. 217.

[60]Franda (1979), p. 230.

the Bangladesh government on the Ganges. Thus, attitudes of the negotiating teams were simply reflecting the political dictums of the ruling regimes.

2.5.2 *Farakka Discussion: Manoeuvring of Politics*

Thus it appears that the Ganges dispute can only be examined by taking into account many above discussed variables. The present section demonstrates how the Ganges issue evolved around the art of out manoeuvring the competitor in domestic political discourse. This section heavily draws on documentations of Indian Parliamentary proceedings to highlight the Indian understanding of the issue and the method to resolve it. Interestingly, in the beginning Awami League did not attribute bad intentions to India for the Farakka barrage; even its 1970 election manifesto did not mention it in poor light:

> The criminal neglect of earlier governments has allowed Farakka Barrage to become a fait accompli resulting in grave and permanent damage to the economy of East Pakistan. Every instrument of foreign policy must be immediately utilized to secure a just solution to this problem.[61]

Later after independence, there was a difference of opinion among members of the government about how to deal with India about Farakka. While speaking to the press, Khondaker Moshtaque suggested that he told Parliament in 1972 that Farakka was a national problem. He continued further that he lost his portfolio of the Minister for Water Resources and Flood Control when he had differed on the Farakka issue with the government handling in 1973 and 1974. Contrarily, it has also been suggested that Moshtaque was prolonging the dispute to further his own political career. This was the thesis of Enayatullah Khan (who was editor of the weekly *Holiday* in 1974 and later became Bangladesh Petroleum Minister in 1978):

> (Moshtaque) was not a hardliner. He was trying to make it big with the issue, trying to make his career out of it. He had almost capitalised on it. …The tussle between Mujib and Moshtaque was not ideological. There was a potential rivalry. Only two people could historically have challenged Mujib: Tajuddin Ahmed (Finance Minister and leader of the Government- in – exile)…and.Moshtaque was the leader of the right-wing faction within Mujib's cabinet. In August 1975, if not earlier, he gave support to the group of majors who staged the coup d'etat in which Mujib and many of his family were killed. For 3 months in 1975 Moshtaque was President of Bangladesh, but in November he was overthrown and later jailed for corruption.[62]

---

[61]Quoted Bangladesh Documents in Ben Crow, et al., "Co-operation and Agreement: 1971–1977," in, p. 102.
[62]Ibid., p. 104.

Towards the late 1974 when Mujib regime was losing popularity for number of reasons, various political leaders were interested to detach themselves from his policies as it happens to politics of any other country. This could be part of distancing from 'unpopular' government and its policies.

At the end of April 1975, Bangladesh's peasant leader, Moulana Bhashani "called on India to dismantle the Farakka Dam… and threatened to organise a protest march of one million people". Bhashani carried out his threat 10 days after the end of the augmentation talks with India. Estimates of the size of the march ranged from one hundred thousand to half a million people. The Indian government prepared for violence and border violations, but the march stopped 6 miles short of the border.

One of the casualties of the coup and counter coups was that during the 1974 accord, when India commissioned the project, both sides had set up three joint teams to observe the effects of the experimental diversions. These three teams began measuring what was happening during the 40 days but joint teams were not successful to submit their joint findings because of disagreement among themselves. In 1977, it was agreed that the JRC should be upgraded to include the relevant ministers from each government. Thus, the JRC was transformed from an official-level forum dominated by technical experts into a political and technical commission. This upgradation of JRC underlines the fact that both governments (Janata government in India and Zia government in Bangladesh) realised that river water was too important an issue to be managed by the bureaucrats only.

When the Janata government struck the agreement with Bangladesh, the Defence Minister Jagjivan Ram also consulted West Bengal government because the immediate affected party in India was West Bengal which needed to be informed about any decision regarding Farakka barrage. The Minister, Ram, visited Calcutta at the end of August 1977 to apprise of West Bengal of the new developments.[63] Making an argument about important role of states which need to be taken into account by the central government, Member of Parliament remarked

> … in our border settlement with Bangladesh are we not keeping the Government of West Bengal constantly in the picture? They are always in the picture and they are a necessary participant in these talks and we are keeping them in the picture. Therefore, there is no question of any State Government being kept out. After all, India consists of States and Union Territories but the question is the people in that area who are immediately affected and their State Governments would certainly come into the picture.[64]

[63]G.K. Reddy, 30/9/77, *The Hindu* 'Agreement on Farakka', and 'Farakka: Dhaka summons Abbas for fresh briefing," the Hindu, 28/9/77.

[64]Demands for Grants, VI/23 Lok Sabha, 9 July 1980.

The news of the 1977 Farakka Agreement had invited strong reactions from Members of Parliament particularly from those belonging to West Bengal. One of the members expressed concern in the following words:

> ...The newspapers have been reporting that India might have agreed to cut down during the lean months the upland flow from the Ganges to less than 20,000 cusecs and this has caused serious concern in Calcutta. Such curtailment to upland supplies to the Hooghly to 50 % of its minimum requirements will, according to the experts, far from effecting a steady improvement of the river, sound the death knell of the Calcutta port. And similar misgivings are being voiced by the people in Bihar. In fact, somebody has pointed out that if this kind of an agreement goes through, then the very purpose of having the Farakka barrage is frustrated and stultified.[65]

Replying to this, the Minister of Defence (Shri Jagjivan Ram) said

> We will resolve our differences, a few cusecs on this side and a few cusecs on that. But what I think is that friendship with the neighbouring countries is of much more value than that. ... My handicap, Sir, is that he has disclosed some figures. Well, opposition does not always mean irresponsibility, especially in matters where the issues are national, where the issues do not even belong to one party or the other. I had the good fortune of negotiating on behalf of the Congress Government and I have the good fortune of negotiating on behalf of the Janata party Government also and I have always held that the Farakka issue is not a party question, but it is a national question and it should be viewed in that light and not as a party question. water is not adequate to meet the requirements of both the countries. Therefore, we will have to share shortages. When we are sharing shortages, certainly there will be discontentment on this side and there will be discontentment on that side as well. But we have to face our people and explain to them the circumstances under which we have accepted the quantity which is not adequate to meet their requirements and they in Bangladesh will have similarity to explain to their people why they have accepted a quantity which is not adequate to meet their requirements. Their difficulty will be more than ours because up till now, even at the time of the minimum flow, they were receiving 100 % water that flow below the Farakka. Now they will be receiving only a percentage of that. We will have to explain to our people that *in the interest of friendly relations between the two countries, we had to agree to share shortages in a manner which can be regarded as friendly*.[66] (emphasis mine)

Thus, the Janata government had to face tough time justifying to agree for a high quantity, which was not high according to Bangladesh expectations but quite high from the recommended quantity for desiltation of the Calcutta port. One of the members in Rajya Sabha made a long statement regarding potential impact being felt by Calcutta as fall out of the new agreed principles.

> ... I do not know whether you have taken any steps—we expected that there would be some statement kept on the Table of the House—to try to investigate that if the flow of the water is diminished, what will be the effect on the Calcutta port. As it is, we are told that there is

[65]Member of Rajya Sabha, VP. Dutt expressed his concern on the debate on the Working of the Ministry of External Affairs, Rajya Sabha C11/17 9th August 1977. pp 76–77.
[66]Ibid., p. 105.

> 112 million tonnes of silt in the Hooghly. The world experts had told us that if there was a flow of 40,000 cusecs at least, then there would be 8 % of desilting annually and in the course of 12 years, on that basis, it would be possible to clear the silt that had already deposited and ships of bigger draught would be able to come to Calcutta, as they did 15 or 20 years back. Now the situation is that if you reduce the flow of water to anything below 35,000 cusecs, there will be a deposit of 2–8 % silt annually, with the result that there can be no question of its clearance in about 10 years. Then the Calcutta port will be doomed, destroyed and ruined for ever. And you are going to enter into an agreement on the dangling of carrots-on more hopes. Bihar and Uttar Pradesh are the States which gave the Janata Party hundred per cent seats in Parliament. This is how you are are going to treat those people in Bihar and Uttar Pradesh. Our people there stand cheated, 92 cheated, cheated. On this assurance, you are going to destroy the destiny of those people for ever. This is the basis of the whole agreement. We are fortunate and happy that agreement has not taken place and I hope that in the days to come, Sir, good sense will prevail and you will go, back on that position and see that the interests—not only of the Calcutta port but of the Gangetic valley—will be protected and preserved.Now, this Farakka question is not only one of sharing of water. It is one of political relations with our neighbouring country as well. The foreign policy of the Government has also to see how far we succeed in having friendly relations with our immediate neighbours.[67]

It was being felt that India is sacrificing its interests in the name of friendly relations:

> After all what are the reasons done to which despite our good intentions, feelings and efforts, we have not been able to establish good relations with our neighbouring countries. I feel that the reasons is that some of the government in our neighbouring countries have not been elected by the people. Today, there are military governments in Pakistan, Bangladesh and Burma. In Sri Lanka also, there is an indirect military rule. It is a fact of history that the military rulers have sought to perpetuate their rule by creating feelings of hatred towards their neighbouring countries or other countries. In Pakistan the military rulers always used hatred towards India as a means to perpetuate their rule. Bangladesh followed suit and the military rulers there want to continue in power by creating the psychology of fear and hatred towards India. In my opinion, we would have to take a two-pronged action in respect of our relations with the neighbouring countries. Firstly, we would have to maintain relations with the people there. Whenever there were democratic governments in these countries, we had friendly relations with them. When there was Mujibur Rahman Government in Bangladesh, when there was a popular government there, our relations with Bangladesh were very friendly. But with the emergence of military rule there, our relations stated deteriorating. Hence, I feel that we would have to establish good relations with the people of our neighbouring countries. It becomes essential for our Government and the people to give moral support to these struggles because so long as these military rules continue, our relations cannot improve with our neighbouring countries, particularly with Bangladesh and Pakistan. When popular governments are restored there, I am fully confident that our relations would improve. With Pakistan and Bangladesh we have got historical relations, blood relations and cultural relations. Only the military dictators have kept us apart. We are all one and we all want to live in peace.[68]

---

[67]Discussion on the Working of the Ministry of External Affairs, Rajya Sabha C11/17 9th August 1977, pp. 103–104.

[68]Member of Parliament, Shri Zainul Basher, Demands for Grants (General), 1985–1986, 8th series III/20 10th April 1985, p. 389.

As explained in the previous chapter, the Ganges Agreement was not renewed further after 1982, Member of Parliament welcomed the decision in the following words:

> Now I can only say this much that I am glad that the present Government has at least done, if nothing else, they have managed to get out of that agreement.... They have not been able to negotiate a new agreement. That is true. But they have got out of that agreement and Bangladesh was demanding that there should be a further 5 year extension of the same agreement which the Government of India has refused to enter into. Now there will be some ad hoc year to year agreement perhaps. But at least it is better than what we were trapped in-that five year agreement which the Janata Government has signed.

The Minister of External Affairs justified his government position regarding the Farakka Agreement:

> Here is a problem and we are greatly concerned because it is not only Calcutta port, but much more that is at stake, U.P is at stake and the entire area which prospectively would have got water from the Ganges and its basin. So, it is not an easy matter for us. We cannot take it lightly and at the same time we know that even Bangladesh cannot take it lightly. Therefore, it will be a test of political will; it will be a test of statesmanship and I am not able to presently envisage a right answer to the question, a right solution, for various reasons, this is the position and therefore we will have to be patient, we will have to be imaginative and we will have to go on with the effort.[69]

Thus above Parliamentary discussions suggest that the sheer desire of the Janata regime to maintain good neighbourly relationship pushed it to have an agreement with Bangladesh. It was driven by calculations of domestic politics to portray an image of friendly regime to its domestic constituency and outside world. But the achievement lies to the fact that at least a negotiation was concluded for 5 years term which was precursor of the 1996 treaty. From Bangladesh political perspective also, when the Awami League decided to negotiate with India on Farakka, it had a reference to make the treaty politically palatable in fierce competitive politics.

Later, the positive development took place during Rajiv Gandhi and General Ershad regime in late 1980s, but domestic political calculations could not let them shape the final result. The Rajiv Gandhi government was occupied to send Indian troops to Sri Lanka, and other issues like the Bofors affair and the crisis in Punjab. Relations between the two governments also deteriorated on account of other factors. Migration of tens of thousands of Chakma refugees from the Hill tracts of Bangladesh into India dominated discussions between the two governments from the early months of 1987. The Chakma migrations, responding to a wave of Bengali settlement supported by army action in the Bangladesh Hill Tracts led to new rounds of accusations between the two foreign ministries. Other issues like Indian soldiers killing Bangladesh civilians and soldiers at the border dominated the

[69]31 March, 1981.

Indo-Bangladesh relationship.[70] Thus, golden opportunity (the new proposal was discussed in the first chapter) to resolve the Ganges river water issue was lost due to domestic political compulsions of both governments.

In 1996, when India and Bangladesh signed the long-term treaty on the Ganges, response of the national parties was different from that of the regional arm. Since late Jyoti Basu was an important architect of the treaty, his party supported it. Though Indian National Congress supported the treaty at the national level in the Indian Parliament, the BJP and the Congress MPs from West Bengal criticised it as a 'sell out' to Bangladesh as it failed to protect the interests of the Calcutta port.[71] Thus the Ganges has been a playground for manoeuvring by politicians against each other for the battle of gaining loyalty of domestic constituency.

2.6 Conclusion

The above discussion suggests that India became closely associated with the Awami League government. This led to a rupture even though briefly between new Bangladesh government and old Indian government of the Congress party after assassination of Mujib. But the change of government in India presented an opportunity to refresh and refashion the relationship. Political expediency pushed change in policy direction of the Desai government in 1977 which paved the way for peaceful interregnum in the stalemate. The above analysis also underlines that the construction of the Farakka was not an intemperate decision by the Indian government with an evil design to harm the neighbour. A number of committees examined the various options of increasing draught of the river before a final call was taken on the Farakka barrage. The repeated demands by West Bengal government and members of parliament forced the central government to go ahead with the project. What emerged out of the Ganges river dispute is that the water issue between neighbours cannot be resolved by taking it as a technical issue of river water sharing because any controversy also reflects social, cultural and historical underpinnings. Potential for manoeuvre becomes high in the domestic politics where history has been fractious. India and Bangladesh negotiated several times before signing a treaty for 30 years. There has always been tussle between India and Bangladesh about preferred way of negotiation-bilateralism or multilateralism. The rationale for preference to a particular way of resolution of river dispute is the subject of next chapter.

[70]Crow and et al., op. cit., p. 207.

[71]Banerji (1999), p. 44.

Annexure

No. 16210

BANGLADESH
and
INDIA

Agreement on sharing of the Ganges waters at Farakka and on augmenting its flows (with schedule). Signed at Dacca on 5 November 1977

Authentic texts: Bengali, Hindi and English.
Registered by Bangladesh on 12 January 1978.

BANGLADESH
et
INDE

Accord relatif au partage des eaux du Gange à Farakka et à l'augmentation de son débit (avec annexe). Signé à Dacca le 5 novembre 1977

Textes authentiques : bengali, hindi et anglais.
Enregistré par le Bangladesh le 12 janvier 1978.

AGREEMENT[1] BETWEEN THE GOVERNMENT OF THE PEOPLE'S REPUBLIC OF BANGLADESH AND THE GOVERNMENT OF THE REPUBLIC OF INDIA ON SHARING OF THE GANGES WATERS AT FARAKKA AND ON AUGMENTING ITS FLOWS

The Government of the People's Republic of Bangladesh and the Government of the Republic of India,

Determined to promote and strengthen their relations of friendship and good neighbourliness,

Inspired by the common desire of promoting the well-being of their peoples,

Being desirous of sharing by mutual agreement the waters of the international rivers flowing through the territories of the two countries and of making the optimum utilisation of the water resources of their region by joint efforts,

Recognising that the need of making an interim arrangement for sharing of the Ganges waters at Farakka in a spirit of mutual accommodation and the need for a solution of the long-term problem of augmenting the flows of the Ganges are in the mutual interests of the peoples ot the two countries,

Being desirous of finding a fair solution of the question before them, without affecting the rights and entitlements of either country other than those covered by this Agreement, or establishing any general principles of law or precedent,

Have agreed as follows:

A. Arrangements for sharing of the waters of the Ganges at Farakka

Article I. The quantum of waters agreed to be released by India to Bangladesh will be at Farakka.

Article II. (i) The sharing between Bangladesh and India of the Ganges waters at Farakka from the 1st January to the 31st May every year will be with reference to the quantum shown in column 2 of the Schedule annexed hereto which is based on 75 percent availability calculated from the recorded flows of the Ganges at Farakka from 1948 to 1973.

(ii) India shall release to Bangladesh waters by 10-day periods in quantum shown in column 4 of the Schedule:

– provided that if the actual availability at Farakka of the Ganges waters during a 10-day period is higher or lower than the quantum shown in column 2 of the Schedule it shall be shared in the proportion applicable to that period;

– provided further that if during a particular 10-day period, the Ganges flows at Farakka come down to such a level that the share of Bangladesh is lower than 80 percent of the value shown in column 4, the release of waters to Bangladesh during that 10-day period shall not fall below 80 percent of the value shown in column 4.

[1] Came into force on 5 November 1977 by signature, in accordance with article XV.

Article III. The waters released to Bangladesh at Farakka under article I shall not be reduced below Farakka except for reasonable uses of waters, not exceeding 200 cusecs, by India between Farakka and the point on the Ganges where both its banks are in Bangladesh.

Article IV. A Committee consisting of the representatives nominated by the two Governments (hereinafter called the Joint Committee) shall be constituted. The Joint Committee shall set up suitable teams at Farakka and Hardinge Bridge to observe and record at Farakka the daily flows below Farakka Barrage and in the Feeder Canal, as well as at Hardinge Bridge.

Article V. The Joint Committee shall decide its own procedure and method of functioning.

Article VI. The Joint Committee shall submit to the two Governments all data collected by it and shall also submit a yearly report to both the Governments.

Article VII. The Joint Committee shall be responsible for implementing the arrangements contained in this part of the Agreement and examining any difficulty arising out of the implementation of the above arrangements and of the operation of Farakka Barrage. Any difference or dispute arising in this regard, if not resolved by the Joint Committee, shall be referred to a panel of an equal number of Bangladeshi and Indian experts nominated by the two Governments. If the difference or dispute still remains unresolved, it shall be referred to the two Governments which shall meet urgently at the appropriate level to resolve it by mutual discussion and failing that by such other arrangements as they may mutually agree upon.

B. Long-term arrangements

Article VIII. The two Governments recognise the need to cooperate with each other in finding a solution to the long-term problem of augmenting the flows of the Ganges during the dry season.

Article IX. The Indo-Bangladesh Joint Rivers Commission established by the two Governments in 1972 shall carry out investigation and study of schemes relating to the augmentation of the dry season flows of the Ganges, proposed or to be proposed by either Government with a view to finding a solution which is economical and feasible. It shall submit its recommendations to the two Governments within a period of three years.

Article X. The two Governments shall consider and agree upon a scheme or schemes, taking into account the recommendations of the Joint Rivers Commission, and take necessary measures to implement it or them as speedily as possible.

Article XI. Any difficulty, difference or dispute arising from or with regard to this part of the Agreement, if not resolved by the Joint Rivers Commission, shall be referred to the two Governments which shall meet urgently at the appropriate level to resolve it by mutual discussion.

C. Review and duration

Article XII. The provisions of this Agreement will be implemented by both Parties in good faith. During the period for which the Agreement continues to be in force in accordance with article XV of the Agreement, the quantum of waters agreed to be released to Bangladesh at Farakka in accordance with this Agreement shall not be reduced.

Article XIII. The Agreement will be reviewed by the two Governments at the expiry of three years from the date of coming into force of this Agreement. Further reviews shall take place six months before the expiry of this Agreement or as may be agreed upon between the two Governments.

Article XIV. The review or reviews referred to in article XIII shall entail consideration of the working, impact, implementation and progress of the arrangements contained in parts A and B of this Agreement.

Article XV. This Agreement shall enter into force upon signature and shall remain in force for a period of 5 years from the date of its coming into force. It may be extended further for a specified period by mutual agreement in the light of the review or reviews referred to in article XIII.

In witness whereof the undersigned, being duly authorised thereto by the respective Governments, have signed this Agreement.

Done in duplicate at Dacca on the 5th November 1977 in the Bengali, Hindi and English languages. In the event of any conflict between the texts, the English text shall prevail.

Rear Admiral
Musharraf Husain Khan
Chief of Naval Staff and Member, President's Council of Advisers in charge of the Ministry of Communications, Flood Control, Water Resources and Power, Government of the People's Republic of Bangladesh

For the Government
of the People's Republic
of Bangladesh

Surjit Singh Barnala
Minister for Agriculture and Irrigation, Government of the Republic of India

For the Government
of the Republic of India

SCHEDULE
(vide *article II (i)*)

SHARING OF WATERS AT FARAKKA BETWEEN THE 1st JANUARY AND THE 31st MAY EVERY YEAR

1 Period	*2* Flow reaching Farakka (based on 75% availability from observed data (1948-73))	*3* Withdrawal by India at Farakka	*4* Release to Bangladesh
	Cusecs	*Cusecs*	*Cusecs*
January			
1-10	98,500	40,000	58,500
11-20	89,750	38,500	51,250
21-31	82,500	35,000	47,500
February			
1-10	79,250	33,000	46,250
11-20	74,000	31,500	42,500
21-28/29	70,000	30,750	39,250
March			
1-10	65,250	26,750	38,500
11-20	63,500	25,500	38,000
21-31	61,000	25,000	36,000
April			
1-10	59,000	24,000	35,000
11-20	55,500	20,750	34,750
21-30	55,000	20,500	34,500
May			
1-10	56,500	21,500	35,000
11-20	59,250	24,000	35,250
21-31	65,500	26,750	38,750

References

Ahmed S (2004) Changing phases of bengali nationalism leading to bengali muslim identity. In: Bangladesh past and present. A.P.H. Publishing Corporation, New Delhi, p 211

Banerji AK (1999) Rivers of joy, rivers of woe: sharing, augmentation and management of water resources in the Ganges-Brahmaputra Basin. S Asian Surv 6(1):39–50

Bhasin AS (ed) (1996) India–Bangladesh relations. In: Documents 1971–1994, vol 1. Siba Exim, Delhi, pp 416–434

Chakraborty S (1974) With Dr. B.C. Roy and other chief ministers (A record up to 1962). Benson's, Calcutta, p 53

Chatterji J (1999) The fashioning of a frontier: the Radcliffe line and Bengal's border landscape 1947–1952. Mod Asian Stud 33(1):196

Dixit (1973) Beginnings of disillusion. p 177

Dixit JN (1999) India gets invoved. In: Liberation and beyond Indo-Bangladesh relations. Konark Publishers, New Delhi, p 49

Franda M (1975) Indo-Bangladesh relations. In: Bangladesh the first decade, 1982. South Asian Publishers, New Delhi in association with Universities Field Staff International Hanover, New Hampshire

Franda (1979) Political realignments. p 230
Ghosh SK (2011) India becomes independent. In: The tragic partition of Bengal. Radical Impression, Kolkatta, pp 387–388
Government of India (Relevant years) Lok Sabha Debates. Lok Sabha Secretariat, New Delhi
Government of India (1961) Ministry of external affairs: a project to save Calcutta. New Delhi
Harun ar Rashid (2003) The Foreshadowing of Bangladesh Bengal Muslim league and Muslim politics 1906–1947. University Press, Dhaka
Hassan S (1989) The India factor in the foreign policy of Bangladesh. In: Kabir MG, Shaukat H (eds) Issues and challenges facing Bangladesh foreign policy. Bangladesh Society of International Societies, Dhaka, pp 44–61
Indira Gandhi (ed) (1972) The refugee influx. In: India and Bangladesh selected speeches and statements March–December 1971. Orient Longman, New Delhi, p 16
Iqbal I (2010) The Bengal delta: ecology, state and social change, 1840–1943. Palgrave Macmillan, London
Jalal A (ed) (1995) The 'Populist' era and its aftermath in India, Pakistan and Bangladesh, 1971 to c. 1993. In: Democracy and authoritarianism in South Asia. Cambridge University Press, Cambridge, p 88
Joya C (1994) Bengal divided Hindu communalism and partition 1932–1947. Cambridge University Press, Cambridge
Kodikara SU (1993) External compulsions of South Asian Politics. Sage, New Delhi, p 134
Maniruzzaman T (2003) The liberation war and its aftermath. In: The Bangladesh revolution and its aftermath. University Press, Dhaka, p 67
Mohsin A (2001) Bangladesh: an uneasy accommodation. In: Alagappa M (ed) Coercion and governance: the declining political role of the military in Asia. Stanford University Press, Stanford, CA
Rashiduzzaman M (1977) Changing political patterns in Bangladesh: internal constraints and external fears. Asian Surv 17(9):797
Ray JK (1986) Political and social tensions in Bangladesh. In: Bahadur K (ed) South Asia in transition conflicts and tensions. Patriot Publishers, New Delhi
Rounaq J (1972) Bangladesh in 1972: nation building in a new state. Asian Surv:208–209
Roy R (1971) The agony of West Bengal: A study in Union-State relation. New Age, Calcutta
Sau SN (ed) (1990) The economics of Calcutta-Haldia Port complex. In: Economic and political weekly, vol 25 (Nos. 18–19), p 1017

Chapter 3
India's Negotiating Strategy on the Ganges: Bilateralism Versus Multilateralism

India is the largest country in South Asia in terms of most of statistical indicators. It constitutes about 75 % of demography, territory and natural resources of all South Asian countries put together. Its GDP is more than three times the combined GDP of the South Asian Association for Regional Cooperation (SAARC). India shares border with all South Asian countries except Afghanistan. In fact, no two South Asian countries have direct borders without mediation of India. All these figures do not inspire confidence among South Asian neighbours towards India. Among South Asian countries, Bangladesh is the one that is surrounded by India from all sides except a narrow stretch in its South east section which is connected with Myanmar. In fact, Bangladesh receives more than 90 % of its river water from India; naturally, even a single whisper about any river water development in India raises Bangladesh's anxieties and immediate reactions.

This chapter elucidates the negotiating process of both India and Bangladesh leading to signing of the treaty on the Ganges. Interestingly, in the context of India and Bangladesh, Ganges water issue looks like a bouquet as it is considered complete only when varieties of flowers are put together; resolution of the Ganges river water dispute has been plausible only when there was a possibility of improvement of relationship in other bilateral matters as well.

The present chapter is structured in the following manner. First part of the chapter discusses India's initial experiences of river water negotiations with Pakistan and Nepal; and its impact on India's future course of interactions with Bangladesh on river water. The second part analyses India and Bangladesh's negotiations on the Ganges. The third section describes Bangladesh's attempt to internationalise the issue. There was an interregnum of multilateralism in bilateral discussion which will be discussed in the fourth section. The last section examines dichotomous preferences of India and Bangladesh's negotiating strategy. The chapter ends with the conclusion.

P. Pandey, *India Bangladesh Domestic Politics*,
DOI 10.1007/978-981-10-2371-2_3

3.1 Evolution of India's Negotiating Strategy on River Waters

India entered into river water negotiations with two immediate neighbours during 1950s. While one was with Pakistan on the western rivers and the other one with Nepal on the eastern rivers. These two negotiations were driven by different contexts and requirements. While India and Pakistan went into negotiation on the Indus river because of existential nature of threat faced by Pakistan; nation-building demands of the nascent nation-state demanded India to negotiate with Nepal on the eastern rivers. Urgency of industrialisation, food security and flood menace necessitated assured power supply, irrigation and flood control, respectively. Since Nepal has huge hydropower potential, India signed two agreements with it. These two negotiations have given India very different experiences.

India and Nepal signed the Agreement on Kosi, the largest tributary of the Ganges in 1954. Another Agreement was on Gandak river in 1959. Both Agreements were basically aimed at flood control, hydroelectrical power generation and irrigation. But soon, both agreements were criticised by the Nepali political parties severely because they believed that India secured disproportionate benefits at its expanse. The criticisms led to revision of the agreements by India and Nepal during 1960s. Nepal gained better control over the rivers after the amendment. It could withdraw water from the Kosi river or any of its tributaries for irrigation or for any other purposes. The revised agreement also added that if any power to be supplied to Nepal is generated in a power house located in Indian territory, the Indian Government would construct the necessary transmission line or lines to a point at the India–Nepal border that would be mutually agreed upon and the tariff rates for electricity to be supplied to Nepal would also be fixed mutually.[1] Since then, these agreements have been continuing but have cast a long shadow on future utilisation of Nepal water by India. On the western front, India has had completely different experiences.

As discussed in the previous chapter about the circumstances under which India and Pakistan was partitioned, so it is not repeated here. Just to recapitulate, when it came to division of the western border, the Radcliffe Commission really found difficult to divide the Upper Bari Doab Canal (UBDC).[2] (The upper portion of the Canal remained in India and the lower portion went to Pakistan). Both countries agreed to maintain pre-partition allocation on the UBDC till 31 March 1948. The agreement worked for the period it was negotiated for but after that east Punjab stopped supplying water to west Punjab. Immediately, both countries engaged into political and legal discussions. Soon an agreement was reached; east Punjab began

[1]Salman and Kishor (2002), pp. 76–77.

[2]The canal system to supply water to Punjab area. With partition, lower portions of the canal system went to West Punjab that was in Pakistan and upper portion came to East Punjab of India which had most of head waters.

releasing water to west Punjab. But Pakistan was interested to have a long-term settlement with India on the Indus for assured supply of water.[3] After lots of hiccups in various rounds of bilateral discussions, both countries, particularly India agreed to accept the services of the World Bank to seek the mutually acceptable solution.

With almost a decade into negotiation, both countries concurred to accept the principles of division of the Indus river intermediated by the World Bank. The Indus River Water treaty was signed in September 1960. According to terms of the treaty, India had to pay 62 million pounds Sterling to Pakistan for new link canals to be constructed for transfer of water. India also agreed to restrict its usage of water on the western rivers for irrigating existing areas and further to develop only 70, 1000 acres of irrigation from these rivers subject to specific conditions. This river water treaty was able to solve the biggest irritating factor that has emerged as a consequence of the partition. It seems that the Indian government had signed the river water treaty expecting other positive externalities coming out of this in bilateral relationship. But later episodes did not satisfy its expectations. An illustration of this can be cited here, within a year of signing of the treaty, Jawaharlal Nehru, Indian Prime Minister made a remark which suggests that whatever hope he had from the Indus was lost "this agreement would open the way to settlement on other problems, but we are where we were".[4]

Though this treaty has been well commended for successful and peaceful negotiation of such a contentious issue all over the world, the Indian government effort was not appreciated by various political parties for making such sacrifices. Members of Parliament blamed the Government for a policy of appeasement and surrender to Pakistan. The impact of this river water dispute on the Ganges will be discussed shortly.

3.1.1 India and Bangladesh Negotiations on Farakka Barrage

Immediately after independence of Bangladesh, both countries began discussing proposition of sharing the Ganges. The question of whether India should construct Farakka barrage or not did not arise because India had already begun construction of the barrage in 1961; so the negotiation was confined to sharing and augmentation of the Ganges.[5] In retrospect, it seems that India was interested to prepare realities

[3]For detailed discussion on Indus river water issue, see Gulati (1973).

[4]Ibid., p. 345.

[5]India has already withdrawn from the Barcelona Convention in 1957 which reads: "no state is allowed to alter the natural conditions of its own territory to the disadvantage of the natural conditions of the territory of a neighbouring state." Quoted in Crow et al. (1995).

on the ground during 1960s so that future negotiations could not compel India to abandon the project. This might be because of the unpleasant experiences of the Indus. In fact, barrage was already ready by 1972.

As discussed in the first chapter that even after couple of meetings, both countries could not have unanimity about the way of augmentation. The Accord in April 1975 allowed India for partial running of the Feeder canal from April 21 to May 31. In August 1975, Bangladesh witnessed coup and domestic political situation changed completely which has been discussed in the previous chapters.

In changed circumstances, Bangladesh put a condition that India stopped diverting water from Farakka first and only then the talks would begin. But India refused to accept this condition. It invited the Government of Bangladesh on February 11, 1976 for talks without preconditions for equitable sharing of the Ganges waters for the lean season of 1976. The government of India released press note on this which reads

> Three joint teams were set up to study the Farakka project to check the navigation in the Hooghly river to check the effects of withdrawal of waters on Bangladesh. The first two teams had submitted agreed reports. The third team could not arrive at a conclusion since the Bangladesh side did not afford opportunities to this team to see if there were any adverse effects on that country.[6]

The text of the statement released by the Ministry of External Affairs reads further

> As the lean season-mid-March to mid-May approaches, when the water discharge in the Ganges drops, the Government of India has offered to discuss the question of utilization of the waters at Farakka. The official response of the Government of Bangladesh is awaited.… India is far the major riparian country for the Ganges waters in terms of the catchment area (98 per cent), ultimate irrigation potential (94.5 %), population of the Ganges basin (94 percent) as compared to Bangladesh. The waters of the Ganges are of vital importance in sustaining the agricultural, industrial and commercial economy of very large areas of India. It is no secret from anyone that the Farakka Barrage Feeder canal has been operating at near or optimum capacity (40,000 cusecs) since June 1975. These withdrawals during a period of abundance of water in the Ganges benefitted not only the port of Calcutta but also Bangladesh itself by diverting some waters which otherwise would have caused flooding. The hue and cry presently being raised in the Bangladesh press, after a lapse of nine months, points to objectives other than those related to the use of Ganges waters at Farakka during the lean months.[7]

This invitation was followed by talks which were held at both technical and governmental levels. India sent two teams to Bangladesh—a technical delegation from April 27 to May 2 and—a "good will delegation" headed by G. Parthasarathy, chairman of the policy Planning Committee of the Indian Ministry of External

[6]"India deprecates Bangladesh's Attitude on Ganges Talks," *The Hindu,* 19 February 1976.

[7]"Ganges Waters: India Refutes Bangla Claim," *The Hindu*, 19 February 1976.

Affairs to Bangladesh from June 18 to 22, 1976 for discussing the Ganges water sharing at Farakka. The Bangladesh delegation was led by Rear Admiral M.H. Khan. As discussed in the first chapter, Khan meeting with Indian Prime Minister lasted a few minutes. The outcome of the meeting was not encouraging. By August 1976, Rear Admiral M.H. Khan had begun taking a tougher stand, declaring that "Bangladeshis would fight to the last and shed our last drop of blood to establish our right, and adding that India wants to cripple us. Unless we stand united and fight out the issue, all our water resources and flood control measures will suffer".[8] By September 1976, both governments could not agree further on the Ganges river water negotiation.

When water talk failed in September, the Bangladesh press went berserk over this. By taking historical perspective into account the Bangladeshi newspaper, the Azad expressed its anguish in following manner:

> Simple minded had thought India's offer of unconditional talks indicated that good sense at last had prevailed upon India and that Farakka dispute will now certainly be settled and a new era of friendship between India and Bangladesh begin….. Knowledgeable, particularly the keen student of the history of Hindu-Muslim relations in India, were wonder struck. Nagged by doubts, they asked themselves: could it really be seen during political rebellions and various movements through the ages that it is not easy for a rival party or side or country to win the heart of the big bosses of New Delhi- in other words, resolve the Hindu-Muslim dispute of the dispute of the subcontinent.. It was again because of this sepoy mutiny, and crush and destroy it….The bitter experience of traversing this futile course was known to Quaid-e-Azam Mohammad Ali Jinnah early in his life. Others to have the same experience were sher-e-Bangla Fazlul Huq, Ali brothers, Mr. Suhrawardy and many more. Therefore, it was asked, would not the historical truth be repeated; would New Delhi accept justice and cheerfully allow Bangladesh to enjoy its rightful share of the Ganges waters? Could the impossible really happen? No, it has not happened; what can never happen has not happened… In short, it is futile to have discussions with India if Bangladesh is to secure its rightful share of the Ganges waters. To achieve this, the people of Bangladesh will have to depend on the international community and ensure that international pressure on India becomes irresistible. Bangladesh thus has raised the Farakka question at the United Nations. This is the way she must forcefully establish her just rights (over the Ganges water).[9]

Another newspaper, The Bangladesh Observer wrote

> …Rear Admiral M.H. Khan found that the invitation was not even remotely aimed at finding a mutually acceptable solution. If anything, the purpose was to reinforce India's unilateral withdrawal of Ganges water by offering unworkable ideas, unilaterally, about the issue." Expressing Bangladesh's frustration the editorial said, "…Bangladesh's attempt at achieving success through bilateral discussions have proved to be a succession of abject failures due to the consistently unresponsive and intransigent attitude of the other side. It has not only rendered Bangladesh's participation in the Joint Rivers Commission absolutely meaningless, but also leaves no alternative to raising the issue in the United Nations… the

[8]Quoted in Kathryn (2000), p. 38.

[9]*The Azad*, 15 September 1976.

> world forum, we hope, can judge the justness and urgency of her case and help solve the problem created arbitrarily by the big neighbour taking an undue advantage of her position as an upper riparian. Since it is vitally connected with the question of Bangladesh's survival. The world body should recognise and discharge its moral obligation in this regard.[10]

There were even claims within Bangladesh that India had registered nine international rivers as "National Rivers" which increased Bangladeshi suspicion of further unilateral actions by India in regards to water sharing.[11] The Lok Sabha (lower house of India Parliament) responded that there had never been such a claim. Thus the atmosphere was vitiated between the two countries in the later part of 1976. After this, President Zia tried to internationalise the Ganges river issue with two objectives—one for domestic constituencies which would enhance its position in people's eyes as it showed to them that his government has stood to India. Another reason was to bring pressure on India.

3.2 Bangladesh's Attempt to Internationalise the Ganges River Water Issue

Seeing no progress on the Ganges river negotiation with India, the Zia government decided to raise the issue at all international fora. Thus, Bangladesh raised the Farakka dispute at various international places beginning with Islamic Foreign Ministers' Conference in Istanbul in the middle of May 1976 and later in the Colombo Summit of the Non-aligned Movement in August 1976 and finally at the UN General Assembly. The issue was brought in the 31st U.N. General Assembly session in November 1976, where India and former Soviet Union strongly protested the inclusion of a bilateral matter in the agenda. Notwithstanding objections, Bangladesh was able to have the issue placed on the agenda but it was not brought before the General Assembly. Rather it was assigned to the Assembly's Special Political Committee where a "consensus resolution" was passed stating that both parties agreed that the situation called for a fair and expeditious resolution.[12]

When the issue was brought before the UN, the Farakka barrage became a sort of legal case to be presented to the international audience; both parties presented their case with utmost conviction. While Bangladesh made a case how Farakka had destroyed all aspects of its lives, India presented a defensive case of how it was crucial for people of this part of the region. The Bangladesh White Paper on Farakka covered all kinds of impacts like ecology, employment, fish production, agriculture, irrigation and industries.

---

[10]*The Bangladesh Observer*, 13 September 1976.

[11]Cited in Joel 2000.

[12]Islam (1987).

The paper began describing the importance of Ganges to the 37 % of the total area of the country which are served by the river, and how one-third of population is dependent on it. According to the paper, the most adversely affected areas were agricultural production because of lack of irrigation and salinity. More than 400,000 acres of land were affected as a result of increased salinity and soil moisture deficiency; more than 400 low lift pumps suffered because of water scarcity and salinity as water level really went down.[13]

The White paper further claims that unavailability of water and deficiency of irrigation facilities led to delay in planting which reduced high-yielding variety yields by 30 %. The reduced flow and increased salinity had damaging effects on the Sundarban (mangrove) forests which was important for maintaining ecology. Sundarban supported employment to many people which they lost as a result of reduced water availability.[14] The paper also addressed the issue of navigation and claimed that the reduced flow of water seriously impeded inland water navigation which is critical in Bangladesh because of the poorly developed transport infrastructure. Not only these, even electric power deficit increased because salinity level of the river water was too high for the tolerance of Goalpara thermal power station and the power station had to be closed for some time. The paper mill and the newsprint mills suffered from lack of power coupled with unusable saline water for processing end products.[15]

In reply, India also presented a strong case in the General Assembly "the city has a population of more than eight million and the operation of Calcutta port directly affects the well-being of other 100 million people in the hinterland, which is much more than the population of Bangladesh. Indeed, the rationale of saving Calcutta city and port impinges upon the economy of as many as 13 states of India as well as the neighbouring countries of Nepal and Bhutan".[16] In reply to the Bangladesh government's White paper, Indian White paper further asserted that Bangladesh has refused to provide the base or benchmark data from which the alleged deteriorations are said to have taken place. This suggests that the facts and figures quoted in support of their case are unfounded. After Farakka, the Ganges continues to flow for 19 km through India, and thereafter flows along the boundary between India and Bangladesh for another 112 km. None of the adverse effects mentioned by the

[13]M.M.Q Mirza, Research: Diversion of the Ganges Water at Farakka and its effects on salinity in Bangladesh, *Environmental Management*, vol. 22, no. 5, pp. 711–722.

[14]From 1981 to 1990, at least two million Muslim Bangladeshi disappeared from the Khulna division of Bangladesh, Ashok Swain, "Displacing the Conflict: Environmental Destruction in Bangladesh & Ethnic Conflict in India," *Journal of Peace Research (Oslo)*, vol. 33, no. 5, December 1996, pp. 959–973.

[15]Bhasin (1996), pp. 416–434.

[16]Statement made by J.S. Mehta in the Special Committee of the General Assembly, *Indian Journal of International Law*, Vol XVI, no. 4, 1976, p. 526.

Bangladesh government have been noticed on the Indian side of the river as a result of withdrawal of waters at Farakka.[17]

Further Indian White Paper gave an illustration of reduced status of the Calcutta port. It stated that the port handled only 7.5 million tons of traffic in 1974–1975 as against 11 million tons in 1964–1965. This decline took place despite "the rapid industrialisation of the hinterland since India's independence in 1947 and increased traffic at other major Indian ports. Calcutta port slipped from second position in cargo-handling operation in 1960–1961 to fifth position in 1970–1971."[18] Moreover the volume of traffic handled in 1974–1975 was much lower than the actual capacity of about 13 million tons a year. Some 50 years ago water drawn from the Hooghly at Palta (28 km north of Calcutta) to serve the needs of Calcutta was almost free from salinity, but in 1975 it became 12 times more saline than permitted by normal health standards. According to the paper, this has led to problems in utilising the waterworks of Palta to their full capacity while the requirements of the city for water have multiplied several times.[19]

In Bangladesh it became an issue of assertiveness so that it can be claimed to local people how the government has stood to big brother and not succumbed to browbeating from India. In this context it is appropriate to quote, in 1978, a minister in the Bangladesh government said that the popular movement in 1976, to which he had been a prominent contributor, "…made Farakka first into a national issue. We tried to transform the issue so that it was no longer just about water… The main purpose (of subsequently going to the UN) was to internationalise the issue in its totality with the other relevant issues. It has helped us in breaking the barriers in the international community. The Bangladesh profile was not very good after 1975 (the coup d' etat)".[20]

Sharing his experience of bringing the Farakka issue to the United Nations, the former foreign Secretary of Bangladesh, Mr. Reaz Rahman stated that getting the Ganges water issue listed in the General Assembly was a big victory because all the big powers of the General Assembly were upper riparian states.[21] Though Bangladesh wanted the issue to be discussed in the main Assembly, the General Assembly referred the issue to the Political Committee where only two persons made interventions, the Bangladesh Foreign Secretary himself and the President of the General Assembly, because, in his view, India had lobbied hard with the member countries. The President of the General Assembly also suggested that both countries should solve the matter bilaterally. This did not dissuade Dhaka from raising the issue elsewhere.

---

[17]India's case on Farakka Barrage, Ministry of External Affairs, Government of India. New Delhi, September 1976. Cited in Bhasin, op. cit., p. 443.

[18]Ibid., p. 437.

[19]Ibid.

[20]Quoted Enayatullah Khan (then Minister of Petroleum), in Ben Crow & et al., op cit, p. 113.

[21]Personal interview of the author with former foreign secretary, Mr. Riaz Rahman on 26 October 2009 at Dhaka.

Bangladesh made a request in the Asian continental meeting of the International Council for Bird Preservation in 1976 that the international agencies should study the means of reducing harmful effects on birds and take into account "the effects already being felt in the Ganges delta as a result of the diversion of water at the Farakka barrage".[22] M.M. Haq, Member, President's Advisory Council in charge of Health, Labour and Social Welfare, raised Farakka issue in the Sixth Asian Labour Ministers Conference of Teheran to illustrate the adverse effect of unilateral withdrawal of Ganges water at Farakka on employment, productivity and mobility of labour in Bangladesh. Despite objections from India, it was agreed that the proposal be transmitted by the conference to the International Labour Organisation (ILO) in Geneva for consideration of the repercussion on the employment, productivity and mobility of labour in Bangladesh arising out of India's decision to withdraw waters from the Farakka.[23]

Bangladesh's policy of internationalising the Ganges issue was, however, not very productive and the deadlock persisted. However a dialogue took place between the two countries following the UN resolution of 1976 at Dhaka. Another discussion took place in January 1977. These discussions were led by Jagjivan Ram, the then Irrigation minister and by Admiral M.H. Khan, Member of the Bangladesh President's Advisory Council in charge of the Ministry of Power, Flood Control. But no forward movement took place on the Ganges. The situation changed only with a change of government in India.

As explained in the first chapter, for the first time, a non-Congress (Janata) government came to power in three decades of Indian independence. The new government enunciated foreign policy options which were different from the previous government. The Janata regime initiated "Beneficial Bilaterialism" regarding neighbours. Beneficial bilaterialism stood for three policies namely personal rapport, economic accommodation, political neutrality and non-interference in internal affairs of neighbours. Immediately within a month of coming into power, the Janata government began discussing river water issue with Bangladesh. On April 18, 1977, after 3 days of negotiations with Bangladesh, the Indian defence minister, Jagjivan Ram announced that an 'understanding' had been reached on the sharing of the Ganges water. Interestingly, Admiral Khan and his team were not ready to accept less than 40,000 cusecs but General Zia overruled them and agreed for 35,000 cusecs. Jagjivan Ram was successful in persuading General Zia to accept his demand of further concession of 500 cusecs. So the deal was pegged at 34,500 cusecs. This was a thorough political deal with hard bargaining from both sides to make the outcome *saleable* in each other's constituency. Here there was no emphasis on technical consideration that only 40,000 cusecs would be sufficient for desilting

[22] "Dacca says Farakka Affects Bird Preservation," *Bangladesh Observer*, 3 October 1976.

[23] "The Farakka raised at Teheran Meet of Asian Ministers," *Bangladesh Observer*, 25 September 1976.

Calcutta port. The only driving force to go for a deal was to give a win-win impression to both parties. This kind of bargaining cannot be negotiated by technical experts and professional diplomats as this requires quick and most importantly momentary decisive decision. It was less than what Bangladesh had expected and certainly less than what Indian engineers had considered the minimum required for the proposed flushing action. For friendship, both parties agreed to climb down from their stated positions.[24] In India, the Defence Minister has overruled his advisors. The generosity showed to Bangladesh by the Janata regime was part of its policy of "Beneficial bilaterialism" regarding neighbours.

As discussed earlier, when negotiation between two neighbours over river water takes place, it is not only about river water but also entails other aspects like overall health of bilateral relationship, bargaining on other issues. This was the case in 1977 Agreement also. When the Janata government signed an Agreement on the Ganges, it went along with improvement in bilateral relationship on other aspects as well. Indian Prime Minister Desai had met Zia at the Commonwealth Heads of Government meeting in London in 1977. An understanding was reached between two governments that India would not support bandit-type attacks on Bangladesh. Indian Prime Minister agreed to "see that no shelter was given to criminal elements from across the border, whatever might have taken place under the previous regime".[25] That is why, non-interference became one of the important foreign policy objectives along with following 'genuine non-alignment' of the Janata government. After the assassination of Mujib ur Rahman, some of his sympathisers had come to India and were involved in raiding Bangladesh and anti-Zia activities from India. The Janata government stopped all support to guerrilla forces who were working in India and were involved against the Zia government.

According to some sources, the Janata government was indiscreet in expelling the guerrilla. There was evidence to suggest that force and compulsions were used in pushing Awami League adversaries of President Zia-ur-Rahman into Bangladesh, who eventually went into the hands of his military regime. The veteran Awami League leader, Phani Majumdar disclosed this to an Indian Journalist 2 years later and explained the experience: "some of our workers had sought refuge in India (after the 1975 crack-down). But since the new government came to power in New Delhi, these workers have been sent back, often at gun point, only to be jailed or tortured by the military rulers". He further added that while doing so in the name of non-interference in internal affairs, the Indian leaders "should be careful that their gestures do not undermine secularism and strengthen obscurantist forces in the countries of the region". Similarly, another Awami guerrilla leader Kader Siddiqui, popularly known as "tiger", was also insulted and compelled to go back.[26]

---

[24]Mehta (2010).

[25]"Zia happy over talks," *The Hindu,* 11 June 1977.

[26]Muni (1979), pp. 417–33.

In fact, General Zia faced lots of domestic troubles including a violent military uprising during the period. Though the mutiny was thwarted, Zia quickly resorted to claims of Indian interference in order to rally popular domestic support. But this time, Zia kept the focus of his attack at West Bengal rather than the Desai government as a whole because of change in circumstances

> (T)hose who take orders from other countries have to leave this country… (T)he West Bengal press have not accepted our rights on the Ganges and were also helping the miscreants. It is significant that the recent incidents took place after the miscreants were allowed to come back and settle down.[27]

Unofficially, details of the *understanding* (of the Agreement) agreed between India and Bangladesh became known within days of the negotiations. This indicated a substantial concession on India's part, which caused considerable concern in Calcutta and West Bengal. Members of Parliament questioned the Janata government about feasibility of survival of the Calcutta port; according to them, the promised amount of water would definitely reduce availability of water for the port. The Minister Jagjivan Ram gave a statement in Parliament clarifying the government's position on Bangladesh and the Ganges issue which is as follows:

> India wanted to decide the question of the Farakka by herself. And, Sir, the question was taken by Bangladesh to the United Nations and there, our friend, mutual friends, friends of India and Bangladesh, all the non-aligned countries, made efforts and emphasised the necessity of India and Bangladesh resolving their differences by themselves. Was it not a challenge to India? Does it require much advocacy and argument to show that in a dispute between two parties, the sympathises go always to the weaker party? Anybody who has dealt with that, will appreciate this aspect. Therefore, it has been our approach to resolve this Farakka question without the intervention of any third party. There are several parties which were very willing to offer their good offices to negotiate between us and resolve our differences. I put it to my friend: will it be very creditable to India to say that we have failed and now we require the services of some third party to resolve out differences? We will resolve our differences ourselves, a few cusecs on this side and a few cusecs on that. But what I think is that friendship with the neighbouring countries is of much more value than that.[28]

There seems to be a consensus, even though misplaced, among Bangladesh's ruling regimes and different political parties that India was forced to agree for water-sharing negotiations in 1977 because Bangladesh had gone to UN in 1976 and that international pressure on India had paid rich dividends.[29] After the end of the term of the Agreement in 1982 again, demands were made regularly to raise the issue in the United Nations and other international fora. The National Council of the ruling Bangladesh Nationalist Party therefore demanded to take the Farakka issue to

---

[27]Quoted in Kathryn (2000), p. 39.

[28]"Discussion on the Working of the Ministry of External Affairs," Rajya Sabha C11/17 9th August 1977, p. 105.

[29]Salman and Uprety (2000). pp. 289–310.

the United Nations.[30] The Vice-president of Bangladesh Muslim League and Chief Advisor of Farakka Sangram Parishad, Nizam Habibuzzaman filed a writ petition in High Court praying for mandatory direction to the Government of Bangladesh to raise and pursue the issue of the Ganges water dispute with India before the United Nations. The petition also demanded that the case should be filed in the International Court of Justice for compensation of damages caused by unilateral diversion of Ganges water by India.[31] Meanwhile protests for restoration of democracy picked up fiercely and the President Ershad was forced to announce an election. The BNP headed by Khaleda Zia formed the government in 1991. When the Bangladesh government did not see any progress on the Ganges negotiation with India, domestic pressure intensified to raise the Ganges issue at international fora. The Progressive Nationalist Party called for a countrywide dawn-to-dusk hartal (strike) on October 10, 1993 in support of their demand to raise this issue at the United Nations.[32] The Jamaat-e-Islami too supported this demand and urged the government to abandon the path of bilateralism.

The Bangladesh Prime Minister Khaleda Zia obliged to these demands and in her address to the UN, pointed out that, "the pledges made by India at the time of commissioning the Farakka Barrage remained unfulfilled".[33] Later, she raised the water-sharing issue in the Commonwealth meeting in October 1993 which according to her had become a "life and death question for the people of the country".[34] Other ministers followed her. At the 27 session of Food and Agricultural Organisation (FAO) in Rome, Agriculture, Irrigation and Water Development Minister, Majidul Haq said that Bangladesh was faced with a grim situation in the desertification of its southwestern region, depletion of fish wealth due to drying of river Padma and its tributaries as a result of withdrawal of water by the upper riparian country, that is, India and that is why, salinity had increased in the coastal areas causing damage to industry and forests in the Sunderbans.[35] Even during the visit of foreign dignitaries to the country, the Bangladesh ministers would raise the Farakka issue. Majidul Haq drew USAID Assistant Administrator for Asia and Near East Ms. Margaret Carpenter's attention towards the problem of the desertification of the southwestern part of Bangladesh due to the adverse effect and uncompromising attitude of India in not releasing desired quantity of water through the Farakka Barrage during the dry season.[36] Another Minister for Posts and Telecommunications Tarqul Islam met China Association for Friendship with Foreign Countries' Vice-President Chen Hau Su in Beijing and sought Chinese help

[30]"BNP wants Farakka Issue Raised at UN," *Morning Sun,* 4 September 1993.

[31]"Writ on Ganges Water Issue Filed in High Court," *Bangladesh Observer*, 12 July 1993.

[32]"BNP'S Farakka stand Hailed," Bangladesh Times, 5 September 1993.

[33]"New Delhi Reacts Sharply to PM's Address to UN," *The Daily Star*, 12 October 1993.

[34]"Khaleda Reiterates Stand on Ganges Water at CHOGM," *Bangladesh Times*, 22 October 1993.

[35]"Majid Raises Farakka Issue at FAO Meet in Rome," *Bangladesh Times,* 14 November 1993.

[36]"Majidul Haq Apprises USAID Official of Farakka Effect," *Bangladesh Times*, 8 February 1994.

in resolving the Ganges water issue.[37] But nothing concrete came out of internationalisation effort this time. Bangladesh's another demand has been the inclusion of Nepal in the Ganges water negotiation since beginning. The option of Nepal exploration is explained below.

3.3 Interregnum of Multilateralism in Bilateral Negotiation

3.3.1 Acceptance of Nepal Option by India

As discussed in the previous chapter, with the change of government in India after assassination of Mrs. Indira Gandhi, Prime Minister Rajiv Gandhi was interested to begin negotiation with Bangladesh afresh to resolve long pending issue of the Ganges river water dispute. Indian Prime Minister met Bangladesh President in October 1985 at the sidelines of the Commonwealth meeting. Both leaders decided to extend cooperation on the Ganges. Following this, a MoU was signed for dry seasons of 1986, 1987 and 1988. Another important aspect of the 1985 MoU was the establishment of Joint Committee of Experts to undertake a joint study to work out a long-term scheme of augmentation of the flows of the Ganges at Farakka. The JCE held all seven meetings from January 1986 to November 1986, with a Ministerial review meeting in August 1986 and a visit to Nepal in October 1986. In between the JCE meetings, there were also a number of meetings of the technical experts assisting the JCE.[38] With the exception of one meeting in 1986, negotiations over water had been exclusively bilateral, that is involving only India and Bangladesh.

Since Indian Prime Minister and Bangladesh President agreed for approaching Nepal jointly, identical letters were sent to Nepal in 1986 which proposed that the JCE would visit Nepal to examine potential of augmenting flows of the Ganges at Farakka through storages in it.[39] The JCE delegation visited Nepal in October 1986. Nepal was interested to know how it would be benefitted by fulfilling India and Bangladesh's demand of data and information for consideration of storage projects in its country for augmenting the flows of the Ganges at Farakka. The joint India–Bangladesh delegation explained that the project had not reached planning stage; that this visit was only a preliminary exploration to see whether projects in Nepal were the right answer to the water-sharing problem at Farakka; that if and when any project planning was undertaken, Nepal would be involved on mutual benefit basis.

[37]"Dhaka seeks China's Help to Solve Ganges Issue," *The Daily Star*, 2 July 1994.

[38]Verghese and Iyer (1993), p. 185.

[39]Rangachari and Iyer, op. cit., pp. 189–190.

Projects could not possibly be undertaken in Nepal without the involvement of that country. Nepal did not find this satisfactory, and suggested that the joint delegation should come back with a clear indication of nature of benefits to Nepal would be. It was agreed between India and Bangladesh that this Nepal venture would be an exploratory joint approach and if a satisfactory response was not forthcoming, the negotiation would be reverted to bilateral nature. It was only on this understanding that India departed from bilateralism; it was not prepared to undertake a prolonged trilateral talks.[40]

The Nepal government seems to be aware of the benefits of a reservoir on the river, for which thousands of acres of land would have to be requisitioned. It is interesting that one of Bangladesh's objections to India's proposal for a Ganges-Brahmaputra link canal has been that this would submerge 20,000 acres of land in Bangladesh which it was reluctant to lose. Therefore, when Bangladesh itself was not prepared to part with its land, expecting the same from Nepal can be called audacious?[41]

Another occasion came in January 1987 at the inauguration of the temporary SAARC secretariat at Kathmandu in Nepal. There was a meeting of the Foreign Ministers of India, Nepal and Bangladesh. At that meeting, it was decided to explore regional way of resolution of the Ganges river water dispute. According to the Bangladeshi analysts, though India had taken the responsibility to provide a detailed report to Nepal outlining future benefits from such projects, it did not do so and dissuaded Prime Minister Rajiv Gandhi from pursuing a trilateral approach to resolve the Farakka issue. Bangladesh's former foreign secretary, Reaz Rahman particularly stressed that, in his view, the Indian bureaucracy had proved to be a hurdle in this case".[42] The retired water secretary put the whole episode in the following way: "The meeting imposed on India the job of preparing a paper on the subject, and this was reluctantly accepted by the Indian Foreign Minister. On return to Delhi, though Indian bureaucracy did take the initiative in proposing a change from bilateralism to regionalism, without a political direction the draft therefore stated the arguments for both courses and sought a political direction. No such political direction came and the paper died a natural death".[43]It is important to mention here that bureaucracy job is limited to prepare procedural details citing advantages and disadvantages of particular policy option; ultimately the decision is taken by the political leaders. Thus, deviation from bilateral path to trilateral exploration option ended like this. After this, there has not been any attempt to explore trilateral or multilateral options. Bangladesh has always been eager to involve Nepal but India strictly wants to follow a bilateral path. The next section analyses rationale for dichotomous preferences.

---

[40]Ibid., pp. 192–193.

[41]Mukhopadhyay (1980), p. 767.

[42]Personal interview with former foreign secretary of Bangladesh at Dhaka in 25 October 2009.

[43]Personal interview with water resource secretary at New Delhi 0n 11 March 2009.

3.4 Bilateralism Versus Multilateralism

As above discussion suggests that India always preferred to have bilateral discussion with neighbours on all contentious issues, only exception has been Kashmir. The Kashmir issue has probably been such a learning experience for the current government and the governments to come, that none of the governments has been interested to go this path. Interestingly, it was India that had brought the Kashmir issue to the United Nations, what follows after that made the Indian government very cautious about involvement of external agency. In fact even during the Indus river water negotiation, Nehru was disinterested in seeking advice from multilateral agencies especially he was not agreeable to third-party adjudication. For example, on October 8, 1950, Prime Minister Nehru wrote to the Pakistani Prime Minister Liaquat Ali Khan suggesting an International Commission consisting of equal number of judges from both countries. Instead of an involvement of multilateral agency, he underlined all possible ways through which issue could be resolved by both parties. Nehru said

> It is true that there is always a possibility of a lack of agreement between the members of the Commission, but if they are judges of the highest standing, they will consider the issues before them in a judicial spirit and are highly likely to come to a unanimous or majority decision. Even if they fail to agree, the area of difference will have been narrowed down by the measure of agreement reached and only the outstanding point or points of difference will remain to be dealt with. The two Governments could then consider the matter afresh, including the question of reference to a third party. To think, *ab initio*, of a third party will lessen the sense of responsibility of the judges and will also be a confession of our continued dependence on others. That would hardly be becoming for proud and self-respecting independent nations.[44]

On the other hand, Pakistan made an attempt to internationalise the Ganges issue even in 1957, when it proposed India to involve the United Nations for technical and advisory services for the development of eastern river systems, but India was of the view that the matter should be explored in a bilateral cooperative way. It seems that by this time India had already begun feeling the heat of multilateral negotiation on the Indus river which had been prolonging for a decade. Since India was not able to commission any projects on the Indus river as it would affect water supplies to West Punjab, Nehru wrote to Mr. Black[45] expressing his concern "at the absence of any progress during this period".[46] Thus India has gone through prolonged process of multilateral negotiation on the Indus.

Another view suggests that the World Bank mediation in the Indus case has resulted into a situation where India had to pay pounds sterling 62 million to Pakistan for replacement works. This experience had probably led to a negative attitude in India towards third-party mediation. In fact Nehru himself felt later that

---

[44]Biswas (1992), pp. 201–209.

[45]Mr. Eugene Black, the President of World Bank of that time.

[46]Quoted in Biswas (1992), p. 207.

the Indus Waters Treaty was generous to Pakistan, while speaking in Lok Sabha, Nehru said "we went pretty far in the Canal Waters Agreement….it was a generous agreement on our part".[47]

India's neighbours in the subcontinent generally prefer to have multilateral discussion to resolve any contentious issues assuming "a multilateral forum will offset India's size".[48] From the Indian standpoint, as the negotiations tend to be lengthy and protracted as it is and would become even more complicated and entangled if three or more countries were to be involved. India, it is argued, fears that smaller countries may join hands to make a "common cause against the bigger country".[49] But India's multilateral engagement has left the bitter taste which is reflected in Indian parliament's discussion. In this context, a debate of Lok Sabha can be quoted here.

> We want durable peace with Pakistan and the basis of that peace should be bilateral. It is a fact, Sir, that there is multi-lateralism in international relations as it has been pointed out… But, at the same time, it is our experience during the last 25 years that multi-lateralism and interference of the third party in the affairs of India and Pakistan have not solved any problems. Rather they have complicated the problems. Even the intervention of the United Nations has not produced any fruitful result so far as durable peace in this sub-continent is concerned. In view of that, if the Government of India takes this attitude that all the disputes, old and new, between India and Pakistan will have to be settled across the table on bilateral basis, I do not think there can be any other more sound policy and there can be any other more reasonable policy.[50]

Even another Member of Parliament strongly objected the idea of bringing Nepal in bilateral discussion. One Member of Parliament expressed this way

> It is a matter of regret that Bangladesh has not responded favourably to our proposals and, instead, has been trying to involve Nepal in the matter, which is essentially a bilateral problem. If we have any problem with Nepal even in the matter of water, we shall settle it with Nepal and in fact we have helped Nepal in executing some projects. But, I do not see why Bangladesh is bent upon making it a trilateral issue. In any case, we must make it absolutely clear to Bangladesh that while we do not want Bangladesh to suffer during dry months, we cannot allow the port of Calcutta to suffer or dry up. They should see reason and agree to an equitable distribution of water. We should also consider and examine—I am suggesting to the Minister—whether the Ganges-Brahmaputra canal can be constructed entirely over our own territory if Bangladesh remains adamant in its attitude.[51]

Even though, the Janata government has agreed to accept the option of exploring the Nepal option through exchange of side letters in 1977, the government reiterated its policy about Nepal option following way

[47]Statement made by Nehru on International Situation, Lok Sabha Debates Second Series, New Delhi: Lok Sabha Secretariat, vol. LVI, 7 to 19 August (Fourteenth Session), 1961, p. 2421–2422.

[48]Iyer (2007).

[49]Iyer (2003), pp. 246–247.

[50]29 November 1972 LXXXII/12 (Pranab Kumar Mukherjee) Motion *RE* International Situation, p. 234.

[51]Discussion on the Working of Ministry of External Affairs Rajya Sabha Debates, vol. CXVIII, NO. 9, 30 April 1981.

> …if in the process of the study of the two proposals, an approach to Nepal was necessary the form and content of the approach would be considered, provided (a) it did not call for Nepal being a party to the existing or proposed arrangements and (b) it did not require Nepal being made a participant in the study in anyway.[52]

In fact, while speaking in the General Assembly of the United Nations on October 4, 1979, Indian Minister of External Affairs, Atal Behari Vajpayee stated that the Farakka Accord vindicated India's faith that complex problems affecting two neighbouring nations should be resolved only "in a spirit of shared sacrifice and mutual accommodation through sincerely motivated bilateral negotiations".[53]

As mentioned earlier, the Ganges issue between India and Bangladesh has always been like a bouquet. River water issue gets affected with the presence of many other bilateral concerns. It would not have been very difficult for two neighbours if negotiation was only about water; it gets embedded or entangled with overall health of the bilateral relationship coupled with domestic political situation. As discussed in the previous chapter, India did not agree for ministerial level discussion with Pakistan even after repeated requests, but began ministerial level river water negotiation with Bangladesh immediately after its independence. Commending the Ganges Treaty, the Indian Prime Minister, Deve Gowda told the Lok Sabha that the visit of his counterpart, Sheikh Hasina Wajed to Delhi to sign the Agreement "has placed our relations on an entirely new footing". This would be "of immense benefit to India in the long term in all areas of bilateral relations including security, trade and other areas".[54] This shows how treaty about river water uplifted the mood in all other spheres as well. In the same vein, the West Bengal chief Minister expressed satisfaction that this river Treaty would pave the way for improvement in other bilateral areas as well. While speaking to the press after the Ganges water treaty, Jyoti Basu made a remark that the pact would bring benefit to West Bengal and with no doubt, Bangladesh could not be without its rewards. He further said

> To our advantage, the option of using the Chittagong Port by our industrialists has opened up significantly so in the context of the State's plans for industrial rejuvenation. Improved relations with Bangladesh could also mean better commerce, trade, easier inter-border transit facilities and the overall development of the entire North-Eastern region.[55]

This explanation holds the ground because India's river relationship with neighbours is not about waters only rather combinations of issues are clubbed. To begin with, Pakistan has river dispute with India on the Indus, it is more interested

---

[52]Quoted in Begum (1988), p. 204.

[53]Muni (1979), p. 65.

[54]Participating in Lok Sabha Debates on 21 February, Fourteenth Session (Part-I) Eleventh Lok Sabha, vol. IX, no. 2, New Delhi: Lok Sabha Secretariat, 1997.

[55]"Agreement not condition: Basu," *The Statesman,* 14 December 1996.

to have control over headwaters of the Indus that lies in Jammu and Kashmir which, according to Pakistan would ensure security of water supply. In this kind of situation, if Indus is discussed at the SAARC level, it will not attract the interests of other members of SAARC because it has nothing to do with them except this kind of occasion can be used for strategic purpose of scoring points either with India or Pakistan depending on political calculations of other countries.

This argument can be extended to other combination of river water disputes in South Asia as well. Bangladesh insists on integrated basin management of the river. The difficulty to go for integrated development of the river can be illustrated with the following example. If integrated development of the Ganges is planned, there will be requirement for correct assessment of availability of water in the river from the source to the end destination. The other variables required will be total record of rains, water coming from melting of ice, suitable location for building reservoirs, assessment for resettlement and rehabilitation cost, each country's water requirement for consumptive as well as industrial usage, further water required for ecological survival of rivers and removal of salinity, etc. This demands sequential thinking to know every minute detail for basin level planning. Thus not only engineers but intersectional group of people like sociologists, anthropologists, hydrologists, geologist, horticulturalists and agricultural scientists are also required to put their heads together to discuss the integrated development of the river. This is desirable for integrated planning of the river. What is the situation at the moment?

As far as India is concerned, water is a classified subject; people have no access to river water record. In fact, how Bangladesh is utilising the present available water of lean period is not made public. Most importantly whether Nepal will be interested to sacrifice its precious land and forest for its neighbours is not known. India and Nepal have already bilateral problems about river water development. No real progress has been made in the Mahakali treaty after two decades of signing in 1996. The Mahakali treaty has proposed Pancheswar Project that plans to irrigate large areas and generate more than 6,400 MW of hydropower to be shared by the two countries. The project was intended to have been built within 8 years, a detailed project report is yet to be prepared. This is also because of different priorities of two governments about utilising water resources. Nepal is in desperate needs of hydropower generation for both internal consumption and exporting electricity. Despite its rich water resources, the country suffers from crippling power cuts because not even 1 % of its hydro potential has been tapped. India's main interests are flood control and irrigation, and if Nepal was able to give them any hydroelectricity, it would be a bonus. As Nepal has become conscious of its water resource potential, Article 126 has been added which reads that "ratification, accession, acceptance and approval of a treaty or agreement" shall require passage by a two-thirds majority of both houses of Parliament present and voting in a joint session in respect of matters concerning "(a) peace and friendship; (b) defence and strategic alliances; (c) the boundaries of Nepal; and (d) natural resources and distribution in the utilisation thereof." The last category especially includes water resource development.[56] In fact, Nepal has been suffering from political uncertainly for almost a decade.

Another important aspect is that India is upstream as well as downstream country. For Nepal and Bhutan, India is downstream but for Bangladesh it is upstream. Here the terms of negotiations change depending upon geographical position of the country. As mentioned earlier, river water issue does not move independent of other bilateral concerns; in that case two neighbours have their own basket of issues along with river water matter. The sharing of costs and benefits is the important bone of contention in the matter of water resource development between India and Nepal. In the case of India and Bangladesh, however, the sharing of the lean season flows, especially of the Ganges, and the augmentation below the Farakka are the main issues.

India–Bhutan relationship is of economic in nature. India helps Bhutan in the construction and operation of hydroelectric projects for the sale of the generated power to India. Though the price of electricity initially negotiated was low that has been corrected by subsequent revisions. The revenues earned by Bhutan by selling electricity to India are substantial for its economy. Now Bhutan is one of the richer countries in South Asia in terms of per capita income. This underlines the argument that though India, Pakistan, Nepal and Bhutan are sharing common rivers, it is India which is in the centre of discussion and the kind of issues India is having with one neighbour is different from that of other neighbours. If another country also becomes involved in a bilateral negotiation, achieving consensus on any project will be tougher. In case of the Indus river with Pakistan, India had to go for third-party negotiation because of bitter political and historical reality. As whatever terms of agreement India had negotiated with Pakistan, the latter would always have suspicion about the former's intentions because of deep distrust. In case of Bangladesh even though it tried to internationalise the issue, it did not fetch her great result except it became public that India has river issue with neighbours on the east as well. Even though two rounds of talks were held between India and Bangladesh on the Ganges after consensus resolution in the UN, nothing was achieved. Whatever was achieved was after the change of government in India and from a government whose sole purpose was to present a different identity to the world in general and neighbours in particular that it was determined to follow friendly and conciliatory approach towards its neighbours. All these have been driven by domestic political rivalries.

There is another perspective which suggests that India has been using its military and economic size to get its will fulfilled on the Farakka issue. "With a military over ten times the size and with thirty times the defence budget of Bangladesh's as well as nuclear capabilities. India's military superiority is hard to dispute, and has played a key role in enabling India to prevail on the Farakka issue".[57] There is a contrary view which presents the issue differently. "However, given India's regional security concerns—epitomized in the recent attacks on Mumbai that were at least in part orchestrated by Pakistan militants–there may be an opportunity for Bangladesh to bring its own political

[56]Verghese (1999), p. 404.

[57]Kathryn (2000).

military power to bear on water issues with India.[58] It further reads, "India's relations with Bangladesh have been most cordial during periods of simultaneous easing of India's "tension-ridden" relationship with Pakistan and China. Conversely, during times of heightened tension between Pakistan and India, Bangladesh may be in a stronger position to seek concessions from India on the Farakka dispute".[59]

As far as exploring United Nations option is concerned, the UN Watercourses Convention of 1997 is the only watercourse convention that has global relevance because it was negotiated by almost every member of the international community including India and Bangladesh. It was adopted by a vote of 103 in favour to 3 against, 27 abstentions [including India and Pakistan]. According to Article 36 of the Convention, "the present Convention shall enter into force on the 19th day following the date of deposit of the thirty-fifth instrument of ratification, acceptance, approval or accession with the secretary-General of the United Nations".[60] The Convention sets forth the general principles and rules governing non-navigational uses of international watercourses in the absence of specific agreements among the states concerned; this also provides guidelines for the negotiation of future agreements. The Convention entered into force on August 17, 2014, after 90 days of the deposit of 35th ratification.

3.5 Conclusion

Thus, India and Bangladesh have had a protracted discussion for almost quarter of a century before signing the treaty on the Ganges in 1996. This period of negotiation was traversed with various ups and downs in bilateral deliberations on the Ganges. The above discussion underlines that Bangladesh's decision to go to international organisations was not driven mainly by the resolution of the Ganges dispute; it was more to demonstrate to the local constituency that the present government would stand to bullying of big upstream neighbour to get its fair share of water assuming this action enhance the prestige of the regime in the eyes of people. But finally the treaty was signed only because of strong political desire from both sides to get rid of the old nagging issue of the Ganges water sharing. As far as preference of negotiating the river water issue is concerned, it seems that the Indus would always be exception keeping India's aversion for multilateralism. Behind the official and political efforts for the treaty, civil society played a major role in the formalisation of the negotiation into the Treaty which will be discussed in the next chapter.

---

[58]Jane Perlez and Salman Masmood, "Pakistan Detains Founder of Group Suspected in Attacks," http.//www.nytimes.com/2008/world/asia/12pstan.html?scp=1&sq=Mumbai%20attacks%20pakistan&st=cse, Retrieved December 11, 2008.

[59]Ibid.

[60]International Water Law Project, http://www.internationalwaterlaw.org/documents/intldocs/watercourse_status.html, accessed on June 19, 2016.

Annexure

TREATY BETWEEN THE GOVERNMENT OF THE PEOPLE'S REPUBLIC OF BANGLADESH AND THE GOVERNMENT OF THE REPUBLIC OF INDIA ON SHARING OF THE GANGA / GANGES WATERS AT FARAKKA

GOVERNMENT OF THE PEOPLE'S REPUBLIC OF BANGLADESH
DHAKA

TREATY BETWEEN THE GOVERNMENT OF THE PEOPLE'S REPUBLIC OF BANGLADESH AND THE GOVERNMENT OF THE REPUBLIC OF INDIA ON SHARING OF THE GANGA/GANGES WATERS AT FARAKKA

THE GOVERNMENT OF THE PEOPLE'S REPUBLIC OF BANGLADESH
AND THE GOVERNMENT OF THE REPUBLIC OF INDIA,

DETERMINED to promote and strengthen their relations of friendship and good neighbourliness,

INSPIRED by the common desire of promoting the well-being of their peoples,

BEING desirous of sharing by mutual agreement the waters of the international rivers flowing through the territories of the two countries and of making the optimum utilisation of the water resources of their region in the fields of flood management, irrigation, river basin development and generation of hydro-power for the mutual benefit of the peoples of the two countries,

RECOGNISING that the need for making an arrangement for sharing of the Ganga/Ganges waters at Farakka in a spirit of mutual accommodation and the need for a solution to the long-term problem of augmenting the flows of the Ganga/Ganges are in the mutual interests of the peoples of the two countries,

BEING desirous of finding a fair and just solution without affecting the rights and entitlements of either country other than those covered by this Treaty, or establishing any general principles of law or precedent,

HAVE AGREED AS FOLLOWS:

ARTICLE - I

The quantum of waters agreed to be released by India to Bangladesh will be at Farakka.

ARTICLE - II

i) The sharing between India and Bangladesh of the Ganga/Ganges waters at Farakka by ten day periods from the 1st January to the 31st May every year will be with reference to the formula at Annexure I and an indicative schedule giving the implications of the sharing arrangement under Annexure I is at Annexure II.

ii) The indicative schedule at Annexure II, as referred to in sub para (i) above, is based on 40 years (1949-1988) 10-day period average availability of water at Farakka. Every effort would be made by the upper riparian to protect flows of water at Farakka as in the 40-years average availability as mentioned above.

iii) In the event flow at Farakka falls below 50,000 cusecs in any 10-day period, the two Governments will enter into immediate consultations to make adjustments on an emergency basis, in accordance with the principles of equity, fair play and no harm to either party.

ARTICLE - III

The waters released to Bangladesh at Farakka under Article I shall not be reduced below Farakka except for reasonable uses of waters, not exceeding 200 cusecs, by India between Farakka and the point on the Ganga/Ganges where both its banks are in Bangladesh.

ARTICLE - IV

A Committee consisting of representatives nominated by the two Governments in equal numbers (hereinafter called the Joint Committee) shall be constituted following the signing of this Treaty. The Joint Committee shall set up suitable teams at Farakka and Hardinge Bridge to observe and record at Farakka the daily flows below Farakka Barrage, in the Feeder Canal, and at the Navigation Lock, as well as at the Hardinge Bridge.

ARTICLE - V

The Joint Committee shall decide its own procedure and method of functioning.

ARTICLE - VI

The Joint Committee shall submit to the two Governments all data collected by it and shall also submit a yearly report to both the Governments. Following submission of the reports the two Governments will meet at appropriate levels to decide upon such further actions as may be needed.

ARTICLE - VII

The Joint Committee shall be responsible for implementing the arrangements contained in this Treaty and examining any difficulty arising out of the implementation of the above arrangements and of the operation of Farakka Barrage. Any difference or dispute arising in this regard, if not resolved by the Joint Committee, shall be referred to the Indo-Bangladesh Joint Rivers Commission. If the difference or dispute still remains unresolved, it shall be referred to the two Governments which shall meet urgently at the appropriate level to resolve it by mutual discussion.

ARTICLE - VIII

The two Governments recognise the need to cooperate with each other in finding a solution to the long-term problem of augmenting the flows of the Ganga/Ganges during the dry season.

ARTICLE - IX

Guided by the principles of equity, fairness and no harm to either party, both the Governments agree to conclude water sharing Treaties/Agreements with regard to other common rivers.

ARTICLE - X

The sharing arrangement under this Treaty shall be reviewed by the two Governments at five years interval or earlier, as required by either party and needed adjustments, based on principles of equity, fairness, and no harm to either party made thereto, if necessary. It would be open to either party to seek the first review after two years to assess the impact and working of the sharing arrangement as contained in this Treaty.

ARTICLE - XI

For the period of this Treaty, in the absence of mutual agreement on adjustments following reviews as mentioned in Article X, India shall release downstream of Farakka Barrage, water at a rate not less than 90% (ninety percent) of Bangladesh's share according to the formula referred to in Article II, until such time as mutually agreed flows are decided upon.

ARTICLE - XII

This Treaty shall enter into force upon signature and shall remain in force for a period of thirty years and it shall be renewable on the basis of mutual consent.

IN WITNESS WHEREOF the undersigned, being duly authorised thereto by the respective Governments, have signed this Treaty.

DONE at New Delhi on 12th December, 1996 in Hindi, Bangla and English languages. In the event of any conflict between the texts, the English text shall prevail.

Signed

(SHEIKH HASINA)
PRIME MINISTER,
PEOPLE'S REPUBLIC OF
BANGLADESH.

Signed

(H. D. DEVE GOWDA)
PRIME MINISTER,
REPUBLIC OF INDIA.

ANNEXURE - I

Availability at Farakka	Share of India	Share of Bangladesh
70,000 cusecs or less	50%	50%
70,000 cusecs -75,000 cusecs	Balance of flow	35,000 cusecs
75,000 cusecs or more	40,000 cusecs	Balance of flow

Subject to the condition that India and Bangladesh each shall receive guaranteed 35,000 cusecs of water in alternate three 10-day periods during the period March 11 to May 10.

ANNEXURE - II

Schedule

(Sharing of waters at Farakka between January 01 and May 31 every year.)

If actual availability corresponds to average flows of the period 1949 to 1988,the implication of the formula in Annex-I for the share of each side is:

Period		Average of total flow 1949-88 (cusecs)	India's share (cusecs)	Bangladesh's share (cusecs)
	01-10	1,07,516	40,000	67,516
January	11-20	97,673	40,000	57,673
	21-31	90,154	40,000	50,154
	01-10	86,323	40,000	46,323
February	11-20	82,859	40,000	42,859
	21-28/29	79,106	40,000	39,106
	01-10	74,419	39,419	35,000
March	11-20	68,931	33,931	35,000 *
	21-31	64,688	35,000 *	29,688
	01-10	63,180	28,180	35,000 *
April	11-20	62,633	35,000 *	27,633
	21-30	60,992	25,992	35,000 *
	01-10	67,351	35,000 *	32,351
May	11-20	73,590	38,590	35,000
	21-31	81,854	40,000	41,854

(* Three ten day periods during which 35,000 cusecs shall be provided.)

References

Begum K (1988) The two proposals for the augmentation of the Ganges and the stalemate. In: Tension over the Farakka Barrage: a Techno-political tangle in South Asia. Stuttgart, Steiner, p 204

Bhasin AS (ed) (1996) Sharing of river waters. In: India–Bangladesh relations. In: Documents 1971–1994, vol 1. Siba Exim, Delhi, pp 416–434

Biswas AK (1992) Indus water treaty: the negotiating process. Water Int 17:201–209

Crow B et al (1995) The decision to build farakka. In: Sharing the Ganges the politics and technology of river development. Sage Publications, New Delhi, p 65

Gulati ND (1973) What the treaty achieves. In: Indus waters treaty an exercise in international mediation. Allied Publishers, New Delhi

Islam MR (1987) The Ganges water dispute: an appraisal of a third party settlement. Asian Surv 27(8):918–934

Iyer RR (2003) Conflict resolution: three river water treaties. In: Water perspectives, issues, concerns. Sage Publications, New Delhi, pp 246–247

Iyer RR (2007) India's water relations with her neighbours. In: USI National. Security Series. K.W. Publishers, New Delhi

Joel MG (2000) The internalisation of disputes over water: the case of Bangladesh and India. Presented on Australasian Political Studies Association Conference, ANU, Canberra, 3–6th October 2000. https://www.google.co.in/search?ix=sea&sourceid=chrome&ie=UTF-8&q=Joel+McGregor%2C+The+Internalisation+of+Dispute+over+water%3A+The+case+of+Bangladesh+and+India. Accessed 28 Mar 2012

Kathryn J (2000) 1975–81: Indo-Pakistani rivalry and Indian party politics. In: Bangladesh, India and Pakistan: international relations and regional tension in South Asia. Macmillan Press, Great Britain, London, Palgrave, p 38

Mehta J (2010) Tryst betrayed: reflections on diplomacy and development. Penguin, Delhi

Mirza MMQ (1998) Research: diversion of the ganges water at Farakka and its effects on salinity in Bangladesh. Environ Manage 22(5):711–722

Mukhopadhyay A (1980) River of contention. Econ Polit Week 15(17):767

Muni SD (1979) India's beneficial bilaterialism in South Asia. India Q 34(4):417–433

Salman MAS, Uprety K, op. cit., and also Nishat A, Faisal IM (2000) An assessment of the institutional mechanisms for water negotiations in the Ganges-Brahmaputra-Meghna system. Int Negot 2(5):289–310

Salman SMA, Kishor U (2002) The Kosi river. In: Conflict and cooperation on South Asia's international rivers a legal perspective. Kluwer Law International, The Hague, pp 76–77

Verghese and Iyer (1993) Indo-Bangladesh talks on 'The Ganges Water Issue'. In: Verghese BG Iyer RR (eds) Harnessing the eastern himalayan rivers regional cooperation in South Asia. Centre for Policy Research with Konark Publishers, New Delhi, p 185

Verghese BG (1999) Waters of hope from vision to reality in Himalaya-Ganges development cooperation. Oxford University Press, New Delhi

Chapter 4
Role of Civil Society in the Ganges Water Negotiations

4.1 Introduction

India and Bangladesh have had a long historical, social and cultural ties. Many familial connections are still maintained through marriages and other means of regular alliances across the border. Even at the governmental level, multiple institutional arrangements have been set up for formal interactions on various subjects. Informally, many groups supported by funding agencies meet intermittently either to build consensus or narrow down differences on issues of bilateral importance.

At the bilateral level, many non-governmental organisations played very critical roles in bringing the neighbours on table. All these organisations have been loosely brought under the genre of civil society. The focus of the chapter is not on conceptual discussion of the civil society; rather on groups which play important roles in bringing people on the table to sort out differences for successful outcome of the discussion. Thus members of Parliament, representatives of political parties and different stakeholders on the issue like the provincial government can be invited to participate in the discussion. In fractious atmosphere, all kinds of possible forum should be harnessed for positive atmospherics and outcomes.

To present an overview, this chapter discusses the definition of civil society and its various dimensions which have been incorporated here. Second, it discusses effort of civil society in resolution of the Ganges river water dispute. The third section evaluates limitations and constraints of civil society in river water dispute. This also explores future potential of civil society and it ends with a conclusion.

Civil society interactions between the two countries have not been a smooth affair, as discussion below will highlight. As the second chapter has already analysed the domestic variables and how they influence interactions of India and Bangladesh, the present chapter explains some constraints which are specifically faced by the civil society.

P. Pandey, *India Bangladesh Domestic Politics*,
DOI 10.1007/978-981-10-2371-2_4

4.2 Defining Civil Society

Civil society is used here as a generic term which includes track-II, track-III, track-IV and track-V groups; non-governmental organisations, subject specific advocacy groups formed to harness resources or for any other common purpose. But broadly track-II level group has been included for the purpose of the present chapter because this segment played an important role in the Ganges river water dispute resolution.

Track-II diplomacy is an organised dialogue in an informal setting where disagreements about each other's government policy can be discussed openly with the possibility of coming to consensus on any contentious issue. In the track-II level dialogue, actors belonging to different walks of life participate which includes former diplomats, academicians, NGO workers, educationists, media persons and even government officials participating in an unofficial capacity.[1] Since meeting is attended by a small group of people, participants are generally unhesitant in expressing their views. Usually, they have a close connection to government, either through the participation of government officials in their personal and private capacities or have direct channels of communication into government circles.[2] This aspect, in the context of the Ganges river water dispute, will be discussed in detail below.

4.2.1 Role of Civil Society in Resolution of the Ganges River Water Dispute

Both India and Bangladesh have large number of civil society groups especially the latter has got distinguished distinction of presence of massive Non-Governmental organisations (NGOs). Visit to Dhaka gives the impression to outsiders that there is an NGO office at every 3 km; thus the city appears as the *city of NGOs*. But, when it comes to interstate relationship particularly in the context of Indo-Bangladesh river water relationship, their presence and visibility is far and few in between. In institutional sense also, the presence and impact of civil society is not visible at governmental level.[3]

Discussions in the previous chapters underline that Indo-Bangladesh Ganges river water negotiations have gone through various ups and downs. At times, Joint River Commission met seven times in a year and in another occasion once in 5 years (mentioned in previous chapters). This is about track-I government level formal interactions. Incidentally, by sheer coincidence, collaborations between civil societies have followed the same trend. This accidental congruence in pattern of

[1] Sewak (2005), p. 37.

[2] Behara et al. (1997), p. 21.

[3] Here civil society has been used mainly in the context of track-II level dialogue.

contacts has not been by design. But in reality, this has been the case. The analysis below would underline this:

With the support of Ford Foundation in 1985–1986, Civil Society groups from three countries—India, Bangladesh and Nepal—began discussion. This track-II initiative involved Centre for Policy Research (CPR) of India, Bangladesh Unnayan Parishad (BUP) of Bangladesh and International Institute of Development Studies (IISD) of Nepal. The purpose of the track two discussions was to evolve a common ground on contentious issues on which both India and Bangladesh could agree. This was the first of its kind for deliberating issues of common concern.

Nine topics were selected for dialogue and three areas were assigned to each country on which it was supposed to prepare an action agenda. The six meetings were held in all three capitals—Delhi, Dhaka and Kathmandu. The action-oriented agenda were prepared. A wide group of people including academicians, politicians and journalists were invited for various sessions. The main objective of this entire exercise was to influence the respective governments to engage in dialogues and to mobilise favourable public opinion for collaborative efforts. The overall theme of these initiatives was "to use water as a vehicle of peace".[4] Thus the fundamental premise was that by solving the water-related issues, political tensions between the countries could be significantly reduced.[5]

This first round of dialogue (1985–1986) did not prove to be successful because members were not willing to form an independent alternative opinion that was different from that of the formal positions of the concerned governments.[6] Its importance lies to an extent that at least it made cross sections of people to sit together at one platform to exchange ideas about issues of India and Bangladesh. Another important point is that ideas which were generated in this session were used for future reference for further discussion. All six meetings were attended by the same group of people. The Indian side did not have any representation from the ruling party, and participants in general had less command over the respective governments in terms of contact and influence.[7] Thus, no concrete output came out of these meetings.

A parallel can be drawn to the result of the formal level initiative. When another MOU was signed between India and Bangladesh in 1985 for three dry seasons of 1986, 1987 and 1988, a Joint Committee of Experts was formed to do a joint study. One of its mandates was to work out long-term schemes for augmentation of flows of the Ganges at Farakka and to identify alternative means of sharing the available river water resources common to both countries for mutual benefit. In spite of seven

---

[4]A. Nishat and I.M. Faisal, "An Assessment of the Institutional Mechanisms for Water Negotiations in the Ganges-Brahmaputra-Meghana System," *International Negotiation*, 5, p. 306.

[5]Ibid.

[6]Ibid.

[7]Ibid.

meetings, the JCE could not finish the work. So, the JCE was granted extension. Even after all these meetings, the JCE could not agree on sharing of river water or modalities of augmentation of flow of the Ganges at Farakka because both parties were not interested to deviate from their earlier held positions. Thus formal as well as informal levels of discussion did not produce any desired result for building consensus on river water. After the end of MoU in 1988, no initiative was taken to resolve the Ganges river issue for a long time. It is interesting coincidence that even at informal level no major initiative was taken to prepare a common ground for resolution of the problem. The result of formal as well as informal levels' interaction has been identical despite the fact that both contacts were not directly feeding each other.

The next significant track-II level interactions were initiated in 1995–1996 period, which was also sponsored by the Ford Foundation. This time, Centre for Policy Dialogue (CPD) of Bangladesh and Centre for Policy Research (CPR) of India arranged meetings on the Indo–Bangladesh relations where various issues were discussed. These meetings were attended by senior and influential politicians and technocrats who had a direct connection to the respective governments. They took messages directly to the governments which made a deep impact. Experiences learnt in the first round of track-II discussion were useful during this round. This initiative helped to bring the two sides closer to each other, particularly the (then) opposition leaders of India and Bangladesh.[8]

The successful discussion at track-II level coincided with the conclusion of the Ganges treaty. Though formally the treaty was signed only on December 12, 1996, the ground work for this had already begun during the concluding part of Khaleda Zia's regime (this has been explained in the first chapter). But it is difficult to say in retrospect that irrespective of change of government in Bangladesh or for that matter in India, treaty on the Ganges would have been signed or not.

Another interesting aspect that needs to be underlined here is the group of people who were involved in the track-II level discussion. In the highly polarised political realities of South Asia, generating bipartisan support has always been difficult in the negotiating process. But the second round of the India–Bangladesh Dialogues has representatives from many political spectrums of both countries.[9] For example, the dialogue participants such as A. Moyeen Khan, Bangladesh's state minister for planning and M. Morshed Khan, who was the special envoy to the prime minister, (part of BNP government of that time) later became front-bench members of the opposition party in Bangladesh's national parliament, the Jatiyo Sangshad. Though the Bangladesh Nationalist Party led by Khaleda Zia vehemently opposed the Ganges treaty arguing that the country's interests had been compromised, it did not seek review of the Agreement when it came to power in 2001. As there has been record in Bangladesh that the two leading parties Awami League and Bangladesh Nationalist Party generally do not agree on anything, the BNP's opposition was

[8]Ibid., p. 308.
[9]Behara (2002), p. 213.

naturally expected; its subtle acceptance was also assumed when no *hartal* on this issue was launched as this is the practice with political parties in Bangladesh. Success for this could be attributed to civil society. As somebody remarked in Dhaka, "two leading ladies of two national parties cannot agree on anything. If the one party signed the treaty with India, the other party would organise strike. If one lady says that sun rises in the east, other will insist that it rises in the West. This is the predicament of our country's politics".[10]

Also, there was an unprecedented movement of the track two participants to the track-I very soon. For instance, from the Indian side, I.K. Gujral, who participated in the dialogue series, subsequently became foreign minister and then the Prime Minister of India. S.A.M.S Kibria from Bangladesh was part of the Dhaka delegation to the first two rounds of the dialogue in Delhi and Dhaka on economic relations, and then became the finance minister of Bangladesh. Thus, it was, according to I.K Gujral, very helpful for the top political leadership to acquire a first-hand understanding of the dispute as well as knowing the outline of alternative policy options for resolving the contentious issues.

In the similar fashion, when negotiation on the Ganges was going on in 1996, the Bangladesh foreign minister Mr. Abdus Samad Azad met Mr. Sitaram Kesri, the Congress president of that time. This might have helped in preparing the background for consensus on broad political spectrum. In fact, there was a sharp criticism of the treaty from the West Bengal unit of the Congress party and members of Parliament. But the central leadership of the Congress (I) welcomed the Ganges treaty. Later the Congress spokesperson described the whole episode this way: "the Bangladesh foreign minister, Mr. Abdus Samad, had called on Mr. Sitaram Kesri a few days ago and the two leaders had discussed the question of sharing the Ganges water. Thereafter, Mr. Kesri and other leaders took up the issue with Mr. H.D. Gowda and Mr. I.K. Gujral and urged them to settle the issue in view of the 25th anniversary of liberation of Bangladesh in which India had played a prominent role".[11] Thus the consensus between the government and the opposition on the important issue brought easy acceptance of policy change among political parties.

4.2.2 Constraints and Limitations of Role of Civil Society

Though historical, political, economic, cultural and of course riverine connections of the two countries have already been explained in the second chapter, that discussion holds importance even for this chapter. Discussion of domestic variables helps analyse the extent of success and failure of civil society's efforts in addressing issues of persistent stalemate such as river water dispute between neighbours.

---

[10]Personal interview with vice-chancellor on condition of anonymity of the prominent university on 25 October 2009 at Dhaka.

[11]"Congress welcomes Ganges Water Treaty," *The Statesman*, 14 December 1996.

Though both countries have large and visible presence of civil societies. They work rather effectively within the national context, keep a keen watch on the government and its delivery system. But when it comes to interstate relationship especially in the context of Indo-Bangladesh Ganges dispute, their impact has been limited. The reason may be that people including members of the civil society have been less critical of post-colonial South Asian states when it comes to interstate relationship. They are guided by the assumption that governments are taking the best decision keeping national interests in mind especially in an issue as vital as water.

Both India and Bangladesh are sensitive about their sovereign status for very different reasons. Bangladesh being a small and downstream country in comparison to India is very conscious about maintaining independent status; at the same time, India is very mindful of keeping its predominant position in South Asia. Keeping all these things in context, civil societies are constrained when it comes to interstate relationship. Also, in case of Bangladesh where politics is fractured and polarised as pro- or anti-India; civil society groups also tread a cautious path because they do not want to be branded as partisan, in this case 'pro-India'.

In case of the Ganges issue, deliberations of a decade before liberation and negotiation of a quarter of a century after independence made opinion and public postures of government, political parties, views of dominant groups known; thus unanimity emerged in Bangladesh that Farakka barrage has ruined all aspects of the country (as discussed in the last chapter about White paper on Farakka in 1976), in this kind of scenario, they are hesitant to take any stand contradictory of the government. Bangladesh press has been complicit in portraying India as an untrustworthy partner and big brother who could devour smaller country like Bangladesh on any pretext; thus going against popular opinion becomes difficult. Maintaining the status quo serves the strategic and political purpose of the government, this assures the domestic constituency that the government is serving national interest. The long-held positions known to people gain such unanimity among the citizens and political parties that presenting the issue with fresh perspective invites trouble to pursuer. This can be explained with an illustration here.

As discussed in the previous chapter that when Bangladesh irrigation minister was preparing consensus on new proposal for river water, this proposal was not even discussed in the cabinet because it was feared that the proponents of the new line could be portrayed as pro-India and anti-Bangladesh. "As ministers or officials within a fragile government (a military regime attempting to make a transition to a government with some democratic legitimacy), they chose to keep the complete package proposal confidential until they could gather more support for it".[12] But when they were brought for the discussion, they thought the plan would not work and that it was more in India's interest than Bangladesh's. Ershad let the proposal slide by.[13] It is important to underscore here that if any Indian or Bangladeshi

[12]Crow et al. (1995), p. 193.

[13]Ibid., p. 198.

proposal is going to solve one country's problem, and reduces another county's difficulties as side-gain, this kind of initiative is abandoned because sharing benefit is not considered complete gain for one party. Even inadvertent munificence is not encouraged. This proposal suffered the problem of this nature.

4.2.2.1 Predominance of Track-II Level Dialogue

As explained above, track-II level discussions are dominated by former diplomats who have been active in office till few months back or in some cases till few years back. Generally, their roles are considered as facilitator but at times this is not the case.

Since diplomats have years of service behind them, they know the intricacies of contentious issues; in their new *avatar* (incarnation), they discuss the matter freely and appreciate the opposite point of view which might not be the case when they were in office. They suggest their counterpart the ways through which they can circumvent the irritating factors coming in the way of resolution of dispute. But sometimes having long exposure become a debilitating factor because they have been so used to government dictates and thinking that they do not budge from rigid stand. In this case, discussion does resemble more like track-I negotiation.

But in most cases they become very liberal as if there have been sudden metamorphosis in the personality. About this, Mr. Verghese has to say, "Once they are no more in official position they can take their own independent position, some of them are genuinely liberal in nature which they behave naturally after retirement".[14]

As discussed above how gradual transformation of the role helps in resolution of the dispute as was the case with Mr. Inder Gujral and others. Since retired diplomats have friends in the government, they can communicate to a government the new thinking or points of emerging consensus. This cannot be done by other people because access to serving foreign officers is not easy in South Asia. There is no platform where researchers can meet serving officers.

4.2.2.2 Lack of Continuous Interactions

Discussions of track-II in a decade 1985–1995 suggest that there were only two talks. This kind of gap does not bring continuity in the deliberations and the process loses the focus required to resolve the contentious matter. As most of the track-II interactions are funded by the funding agencies; they dispense funds according to preference of policy directions of that particular year of the organisation. Getting financial support for continuous collaborations becomes a tough job. India-Bangladesh discussions suffer from this problem most often.

---

[14]Personal interview with the author on 15 May 2012 at New Delhi.

4.2.2.3 Predominance of Creative Ideas or Above the Ground Ideas

Generally, civil society groups are full of creative and imaginative ideas for resolution of the issue. There have been suggestions for integrated river development or basin wide development of the Ganges. They are excellent ideas. But the situation on the ground does not allow the execution of these noble ideas. Mr. Ramaswamy Iyer writes that the advocacy of regionalism tends to become dogmatic. Groups or individuals who suggest this kind of ideas begin to think that they owe the problem and it is their way of handling the situation is the best way to remove the problem.

> Some problems and issues are best dealt with on a national or local basis; some call for cooperation between the two countries or units; and others demand a regional approach. The circumstances vary from case to case, and in each case the most appropriate route needs to be followed. Rigid bilaterialism such as that adopted by the government of India may be unwise and unduly self-limiting, but a dogmatic advocacy of regionalism can unnecessarily complicate simple issues and render resolution more difficult.[15]

This kind of integrated planning of river should come with presentation of desirable results along with an action-plan which could be executed effectively within particular time frame. Getting consensus between two provinces in India is so difficult, generating consensus on any project from two or three countries are going to be tedious affair. As it has already been discussed that river issue is not about water only, it concerns many bilateral issues which make resolution not an easy affair. Persistent effort to create public opinion and depoliticisation of certain aspects of public good will go a long way in sustainable peace.

4.2.2.4 Thinks Tanks Specific Goals

Presently, bilateral institutional interactions are growing between India and Bangladesh at formal as well as informal level. During the Ganges river water dispute, mainly two important track-II level interactions took place. Now many Delhi-based think tanks have institutional collaboration with Dhaka-based counterparts. Institute for Defence Studies and Analysis (IDSA) had institutional collaboration with Bangladesh Institute of International and Strategic Studies (BIISS) (2011), Observer Research Foundation has with Bangladesh Enterprise Institute (BEI) (2011).[16] Though these are bilateral institutional collaborations, there are still some other partnerships which are funded by funding agencies which hold discussions periodically. There was a collaboration between the Kunzru Centre for Defence Studies and Research and Centre for Foreign Affairs Studies supported by Asia Foundation (India and Bangladesh). This dialogue went on from 2009 to 2011

---

[15] Iyer R (2002), p. 281.

[16] According to the website which suggests institutional collaborations with above instates. http://www.idsa.in/internationalinteractions, http://www.observerindia.com/cms/sites/orfonline/html/partnership/institution.html, accessed on 3 July 2012.

and came out with list of recommendations, which were submitted to both countries' foreign ministry.[17] In September 2011, Jamia Milia Islamia and Dhaka University also signed MoU. Under this programme, a workshop was organised consisting of ten young scholars from India and Bangladesh each. Participation in the course provided a unique opportunity to scholars to go through all aspects of Indo–Bangladesh relationship intensely for ten days apart from giving a unique platform for developing interpersonal relationship among themselves. But in some cases, interactions die in the absence of financial support.

4.3 Future Potential of Role of Civil Society

These days, more informal interactions at track-III level are emerging between the two countries. This is being used by NGOs and social activists. Their focus is on contemporary policy issues. They are composed of individuals and groups looking across national frontiers and states, aiming to build constituencies for peace which can question conventional practices and beliefs and present alternatives to official government positions. They do not have access to government agencies but aim to change public attitudes and mobilise public pressure on their respective governments to resolve differences and disputes.[18] There are groups in Manipur (a Northeast Indian state) and Bangladesh which are against construction of Tipaimukh dam,[19] these groups are making a joint cause. They have organised meetings in northeast India as well as Dhaka.

As the role of government and opposition is interchangeable, it is essential to maintain relationship with both of them. If not more at least equal level of interactions with opposition members are desirable. As mentioned earlier, how involvement of opposition members in the track-II level discussion was effective in the Ganges water treaty, this leads to continuity.

4.4 Conclusion

Thus civil society played an important role at the critical juncture of the Ganges River water dispute. As above discussions suggest that civil societies suffer from serious limitations in terms of funding, access to Ministry of foreign affairs and fractured domestic political discourse.

---

[17]Interestingly, when 51 point Joint Development was issued when Prime Minister Manmohan Singh made a visit to Bangladesh, some of the recommendations got reflected from track-II level discussion. Author has been part of this group.

[18]Behera et al (1997) p. 21.

[19]Tipaimukh dam is a proposed embankment dam on the river Barak in Manipur state in India.

Presently track-II level interactions are mostly confined to Delhi and Dhaka. Involvement of provincial groups and people presents a far more real picture of the ground which helps in bringing real possibilities and gauging public opinion. This further reduces irritants between the two neighbours. Seven states of Indian northeast and West Bengal have close connections with Bangladesh which need to be exploited by the civil society groups. Whenever relevant for the cause, civil society interactions must include members of political parties as well as members of Parliament of both countries. This would take relationship to new level and also helps avoiding the kind of embarrassment India faced during Prime Minister Manmohan's visit to Bangladesh in September 2011 when India could not sign the Teesta Agreement which was supposed to be high on agenda before the visit. The Indo–Bangladesh future relationship demands inclusion of these new imperatives. Potential of civil society can be exploited to meet emerging challenges like climate change and other vicious issues.

References

Behera NC (2002) Forging new solidarities: non-official dialogues. In: Mekenkamp M, Van Tongeren P, Van de Veen H (eds) Searching for peace in Central and South Asia: an overview of conflict prevention and peace building activities. Lynne Rienner Publishers, London, p 213

Behera NC, Evans PM, Rizvi G (1997) Beyond boundaries: a report on the state of non-official dialogues on peace, security & cooperation in South Asia University of Toronto-York University Joint Centre for Asia Pacific Studies, Ontario, p 21

Crow B et al (1995) The rise and fall of new initiatives: 1983–1987. In: Sharing the ganges the politics and technology of river development. Sage Publications, New Delhi, p 193

Iyer R (2002) Water-related conflicts: factors, aspects and issues. p 281

Nishat A, Faisal IM (2000) An assessment of the institutional mechanisms for water negotiations in the Ganges-Brahmaputra-Meghna system. Int Nego 5(2):289–310

Sewak M (2005) Track two diplomacy: possibilities and challenges. In: Multi-track diplomacy between india and pakistan: a conceptual framework for sustainable security. Manohar, New Delhi, p 37

Chapter 5
The Ganges Treaty: Future Directions

The Ganges river water treaty has completed nearly two of its three decades' terms. This is no small achievement considering previous records of all Agreements and MoUs taken together lasted for only 11 years. Though this gives a sense of accomplishment and satisfaction for uninterrupted functioning of the Treaty, some old matters are still troublesome and other new issues are emerging which will have important bearing on survival and further extension of the Ganges treaty after 2026 when it finishes its three decades terms. Here, the focus is to analyse variables which have made the operation of the Treaty uninterrupted possible. This is also to address prospective issues, which might have seemed marginal to the treaty at that moment, are already showing signs of arrival and are going to occupy forefront in the near future. Climate change phenomenon looks more real today than previously thought. It is estimated to affect the Himalayan glaciers which are the main sources of the dry season flow for the Ganges. Besides, there are political issues which are going to affect future extension of the treaty. India is going through a transition from a 'unitary state with federal features' to federal units becoming assertive about their interests. To discuss this, the chapter is organised in the following way. In the beginning, the chapter evaluates the operation of the Ganges treaty and also functioning of the Joint River Commission. Next issue being discussed is the impact of climate change on the Ganges river water treaty. The third section examines the changing dimensions of Indian federalism and its impact on Indo-Bangladesh bilateral relationship. The chapter ends with the conclusion.

5.1 Experience of 15 Years of the Ganges Water Treaty

Though the first chapter has already discussed the operational aspect of the Ganges treaty to an extent, here the focus is limited to analyse how the Treaty has managed to achieve immunity in polarised domestic politics. The data presented below will help in understanding this aspect. The table presented below are from 1989–1995, 1997–2000 and another from 2008–2016 of river water being released from India.

P. Pandey, *India Bangladesh Domestic Politics*,
DOI 10.1007/978-981-10-2371-2_5

Period	Flows during 1989–1995 (cusecs)	Treaty	1997	1998	1999	2000	2008	2009	2010	2011	2012	2013	2014	2015	2016
January															
01–10	53619	67516	70821	182263	95934	94975	95883	96414	86720	96179	136486	94211	110853	98528	72335
11–20	47726	57673	55788	154292	85728	79568	81785	90654	77204	88036	111240	82897	107311	102330	62788
21–31	40781	50154	50045	118313	81480	62238	81813	86974	75048	78867	115846	80129	117003	93141	62028
February															
01–10	33417	46323	48430	88363	64873	55903	85113	81650	76266	66516	111872	84790	116284	85071	59474
11–20	26868	42859	38319	73584	61760	53292	73520	73272	68108	64189	102583	81241	110506	81017	57640
21–31	24559	39106	25689	54242	53185	46909	67446	64858	64824	65086	92815	99624	102692	72429	53730
March															
01–10	22868	35000	23291	46686	41600	44573	58872	61226	60742	63373	79208	103879	101999	67058	50838
11–20	19573	35000	19930	40192	35683	39320	59324	56826	57276	57466	65342	92492	106243	79644	50464
21–31	17516	29688	13823	38685	33892	35509	53310	56114	51735	60436	67772	77757	92366	84061	50606
April															
01–10	17177	35000	17857	43960	35376	37026	56474	55211	53696	57848	69192	70727	85994	100892	51573
11–20	19578	27633	24559	53241	30725	35528	52519	54219	52191	59442	80864	68302	80850	94325	53282
21–31	21759	35000	27695	53627	34738	41535	59532	56329	53652	56819	72927	67071	74949	95618	53802
May															
01–10	23467	32351	26578	74886	37672	56126	55353	57446	55791	57800	68059	75117	71523	104165	51648
11–20	32228	35000	26279	92039	41818	50344	64528	59310	55268	65256	64654	72868	71919	106203	52311
21–31	50410	41854	27520	84965	48716	85274	67894	65780	65350	83020	65915	87135	80549	111193	59858
Period	Flows during 1989–1995 (cusecs)	Treaty	1997	1998	1999	2000	2008	2009	2010	2011	2012	2013	2014	2015	2016

(continued)

(continued)

Period	Flows during 1989–1995 (cusecs)	Treaty	1997	1998	1999	2000	2008	2009	2010	2011	2012	2013	2014	2015	2016
January															
01–10	53619	67516	70821	182263	95934	94975	95883	96414	86720	96179	136486	94211	110853	98528	72335
11–20	47726	57673	55788	154292	85728	79568	81785	90654	77204	88036	111240	82897	107311	102330	62788
21–31	40781	50154	50045	118313	81480	62238	81813	86974	75048	78867	115846	80129	117003	93141	62028
February															
01–10	33417	46323	48430	88363	64873	55903	85113	81650	76266	66516	111872	84790	116284	85071	59474
11–20	26868	42859	38319	73584	61760	53292	73520	73272	68108	64189	102583	81241	110506	81017	57640
21–31	24559	39106	25689	54242	53185	46909	67446	64858	64824	65086	92815	99624	102692	72429	53730
March															
01–10	22868	35000	23291	46686	41600	44573	58872	61226	60742	63373	79208	103879	101999	67058	50838
11–20	19573	35000	19930	40192	35683	39320	59324	56826	57276	57466	65342	92492	106243	79644	50464
21–31	17516	29688	13823	38685	33892	35509	53310	56114	51735	60436	67772	77757	92366	84061	50606
April															
01–10	17177	35000	17857	43960	35376	37026	56474	55211	53696	57848	69192	70727	85994	100892	51573
11–20	19578	27633	24559	53241	30725	35528	52519	54219	52191	59442	80864	68302	80850	94325	53282
21–31	21759	35000	27695	53627	34738	41535	59532	56329	53652	56819	72927	67071	74949	95618	53802
May															
01–10	23467	32351	26578	74886	37672	56126	55353	57446	55791	57800	68059	75117	71523	104165	51648
11–20	32228	35000	26279	92039	41818	50344	64528	59310	55268	65256	64654	72868	71919	106203	52311
21–31	50410	41854	27520	84965	48716	85274	67894	65780	65350	83020	65915	87135	80549	111193	59858

Thus, the above table draws the picture which suggests that in each 10-day period water received by Bangladesh from Farakka has been more in post 1996 period in comparison to 1989–1995 except in one 10-day period in March 1997. This means people are in better position post 1996 than previously without treaty.[1] This probably explains BNP's quietness about demand for review of the Treaty provisions when it came to power in 2001.[2] When people are really benefitting from new arrangement, it is difficult to revoke the Treaty. Another noticeable aspect is that the data about river water received by Bangladesh from India has been released by the Awami League government only. During the Bangladesh Nationalist Party (BNP) regime, the data has not been released. This reinforces that the partisan approach plays an important role in the bilateral relationship as the previous analysis suggests. Different Political factions and interest groups continue to exploit any dispute with their 'powerful' neighbour India to gain domestic political leverage. Leaders are aware that cooperation with India can be viewed as compromising national sovereignty and interest.[3] But if the BNP is not making sort of comments like injustice and "selling out interests of the country (usual remarks reserved for the Awami League for concluding any agreement with India as discussed in the first chapter), this implies subtle acceptance if not open embrace of the provisions of the treaty. There have been three general elections in Bangladesh in 2001, 2008 and 2014 (this election was boycotted by the BNP) after the 1996 Treaty but the Ganges river issue was not made an electoral issue by political parties in general and by the BNP in particular[4] and its allies even though India as such was an issue. The above data suggests that India has largely been releasing water to Bangladesh according to provisions of the treaty. In some 10-day periods, water releases are less than scheduled but overall compliance has been maintained.

As far as sharing and releasing of water data is concerned, India has taken a decision that river water data is a classified subject maintaining an archaic policy even at the cost of being charged that it is not releasing water to Bangladesh according to provisions of the Treaty. Even for strategic reasons, it makes sense to make public figures of water released to Bangladesh but Indian government is more comfortable with an old policy. On being questioned about declassification of river water data, one of the senior water resource ministry officials has to say, "this issue of declassification has been discussed at the Prime Minister level also but nothing has happened. Means nobody is taking decision on this issue".[5] So the status quo is

[1]After 1988, they failed to reach an agreement due to decreasing availability of water at Farakkabecause of the upstream withdrawals in Northern India. (Swain 2012), p 72–87.

[2]Interestingly, in all conferences related to Indo-Bangladesh, it is suggested that the BNP government when it was in power in 2001–2006 asked for review of the treaty.

[3]Elhance (1999), Salman and Kishor (2002).

[4]The Bangladesh Nationalist Party (BNP) boycotted the 2014 election.

[5]The author interview with government official, New Delhi, May 2012.

maintained. In 2013, India announced a new policy called Hydro-meteorological Data Dissemination Policy. This classifies data into three categories: 1. Indus basin and tributaries discharging into Pakistan; 2. Ganges-Brahmaputra-Meghna basin and other rivers and their tributaries discharging into Bangladesh/Myanmar; 3. remaining other rivers and their tributaries

The data of category 1 and 2 are classified, whereas category 3 data is unclassified. In certain conditions, one can ask for access to classified data. But the conditions imposed for access to classified data are tedious; a classified data can be released for a specific study and it is non-transferable. The Ministry of Water Resources is made the final authority to decide whether classified data can be released or not. But being acquainted with Indian bureaucratic set up, walking through a web of bureaucratic machinery for permission would be a well-nigh impossible.

5.2 Operations of Joint Rivers Commission (JRC)

An important institution that has been set up to maintain a regular interaction for development of common river waters shared between India and Bangladesh is the Joint Rivers Commission (JRC). The Statute of the Joint Rivers Commission was signed on November 24, 1972 though the first meeting was held in June 1972. The JRC has a Member-level Standing Committee and Local Committees to address local issues but the final approval for a proposed solution needs to come from the respective governments. In 1977, the two countries agreed to upgrade the chairmanship of the Commission to a ministerial level during the visit of the President of Bangladesh to India in December 1977, a few weeks after the signature of the 1977 Agreement. This gave the Commission more political prominence and authority in bilateral discussion. The 1977 Agreement regenerated the Commission as the last meeting was held only in June 1975 which was immediately after the expiry of the Partial Accord.

Political considerations are decisive factors in convening meeting of the JRC. According to statute of the JRC, it should meet four times a year but sometimes it does not meet even once a year. The JRC held six meetings between June 1974 and June 1975; during those meetings, the main issues of discussion were the augmentation of the flow of the Ganges. The JRC held its 31st meeting in Dhaka in June 1990; this meeting happened to be last meeting of the Commission for a long time because next meeting took place only after 7 years in 1997. Again JRC held its 33rd round of meeting in April 1999. Interestingly till 1988, the JRC used to discuss all rivers shared between India and Bangladesh including the Ganges, the Brahmaputra and the Barak. Later, discussion became focussed primarily on the Ganges.

Record of JRC Meetings

Year	Place	Round of meetings
25–26 June, 1972	New Delhi	1st Meeting
28–30 September, 1972	Dhaka	2nd Meeting
11–13 December 1972	New Delhi	3rd Meeting
29–31 March, 1973	Dhaka	4th Meeting
19–21 July, 1973	New Delhi	5th Meeting
8–10 November 1973	Dhaka	6th Meeting
28 February–2 March 1974	New Delhi	7th Meeting
6–12 June 1974	Dhaka	8th Meeting
17 July 1974	Calcutta	9th Meeting
29 August–2 September 1974	Dhaka	10th Meeting
28 September–4 October 1974 16–20 November 1974	New Delhi Dhaka	11th Meeting
31 December 1974–7 January 1975 Minister Level Meeting(Sernibat-Jagjivan) 16–18 April 1975	New Delhi Dhaka	12th Meeting
19–21 June 1975	Dhaka	13th Meeting
20–24 January 1978	Dhaka	14th Meeting
5–11 July 1978	New Delhi	15th Meeting
6 November & 8–10 December 1978	Dhaka	16th Meeting
8–12 May & 16–20 Nov 1979	New Delhi Dhaka	17th Meeting
26–29 February 1980 26–29 April 1980	New Delhi New Delhi	18th Meeting
09–11 July 1980	Dhaka	19th Meeting
29 August–1 September 1980	New Delhi	20th Meeting
26–28 February 1982	Dhaka	21st Meeting
28 August 1982	Dhaka	22nd Meeting
22–24 December 1982	New Delhi	23rd Meeting
2–4 February 1983	Dhaka	24th Meeting
18–20 July 1983	Dhaka	25th Meeting
13–15 February 1984 30–31 March 1984	New Delhi Dhaka	26th Meeting
14–16 December 1984	New Delhi	27th Meeting
2–4 June 1985	New Dhaka	28th Meeting
11 May 1987	New Delhi	29th Meeting
18–19 April	New Delhi	30th Meeting
21–22 June 1990	Dhaka	31st Meeting
19–20 July 1997	Dhaka	32nd Meeting
10 April 1999	New Delhi	33rd Meeting

(continued)

(continued)

Year	Place	Round of meetings
12–13 January 2001	Dhaka	34th Meeting
29–30 September 2003	New Delhi	35th Meeting
19–21 September 2005	Dhaka	36th Meeting
17–20 March 2010	New Delhi	37th Meeting

According to Article IV of the 1996 Treaty, a Joint Committee has been set up. This Committee has joint teams for discharge observations of daily flows at Farakka in the Ganges, in the Feeder canal and in the navigations lock from January 1, 1997. A Joint teams has also been set up at Hardinge Bridge in Bangladesh for joint observations. After the joint observations are completed in May every year, the Annual Report is submitted incorporating the data. Report is not made public so people have little idea about its functions.

Generally, it is claimed that when the BNP government came to power in 2001, it asked for review of the treaty in the 36th round of JRC meeting in 2005. But discussion of the minutes of the Joint River Commission meetings does not support this view. This is the excerpt from 36th round of the JRC held between the BNP and Congress government which are as follows:

> The Commission reviewed the progress of implementation of the sharing arrangements under the 1996 Indo-Bangladesh Treaty on Sharing of Ganges/Ganges waters at Farakka. The Bangladesh side stated that during the dry season of 2005 it received lesser quantity of Ganges waters as its share compared to the quantum shown in the indicative schedule of the Treaty in different 10 day periods. The Bangladesh side referred to provisions of Article-II (ii) of the 1996 Ganges Waters Treaty and requested the Indian side to do the needful to ensure the due shares of both countries. The Bangladesh considered that a sustainable solution to the problem can be found only through augmenting the dry season Ganges flows at Farakka. According to Article-VIII of the 1996 treaty, the two Governments have already recognized this as well as the need to cooperate with each other in finding a solution to the long term problem of augmenting the flows of the Ganges/Ganges during the dry season. The Bangladesh side stated that India and Nepal were undertaking steps to implement storage reservoirs on Mahakali and Saptkosi which Bangladesh had originally proposed in 1985…

It reads further

> The Indian side pointed out that Article-II (i) of the Treaty specifies that sharing of water in any given year will be with reference to the formula as set out in Annexure-1 of the Treaty and this has indeed been done. Article-II (ii) requires that 'every effort' will be made by the upper riparian to protect flows as set out in Annexure-II. The Indian side stated that the average receipts in the past nine years by Bangladesh for each of the specified 10-day periods equalled or exceeded the flows indicated in Annexure-II except for one 10-day period of March. On the other hand, the receipts by India have been less than the indicative flow for the entire period of February and March two 10-day periods of May. It was stated by the Indian side that. This clearly established that India had made 'every effort' as enjoined as Article-II (ii)…The Indian side pointed out that *if they so desire Bangladesh could seek a review of the sharing arrangements as provided for in Article–X*... referring to

> the remarks of the Bangladesh side wherein the cooperation India and Nepal on harnessing of water resources was raised, the Indian side emphasised that the JRC is a bilateral forum between India and Bangladesh and this forum cannot be diluted by raising issues pertaining to any other country.[6] (emphasis mine).

Thus even during the two JRC meetings of the BNP government, no review of functioning of the treaty was demanded. Another aspect is that Bangladesh has been demanding that 40 years average should be maintained, in the 37th JRC meeting "the Indian side clarified that the schedule given at Annexure-II of the Treaty is only indicative. Thus it would match only in a year when the actual availability of water at the Farakka corresponds to the long term average flows of 40 years. Further, a close look of jointly observed flows of last 13 years shows that Bangladesh side has got, in fact, more water (16.34 bcm) than its share (15.03 bcm) as per the indicative schedule of the treaty".[7]

Political considerations mostly decide the meeting. After a gap of 5 years, JRC met in March 2010. A senior officer in the Ministry of Water Resources described the situation of last five years this way. "There was a demand by Bangladesh for JRC meeting but Indian Water resources ministry kept telling Bangladeshis that we are preparing for a meeting, the moment it is ready we will meet", he further adds, "for instance, at the technical level meeting in 2003, anti-erosion activity was discussed but no decision was taken because of differences about how to do this. Good political atmosphere really plays an important role. If the brief comes from the political level, the mechanism is found for solution. Again, in August 2007, secretary level meeting took place for taking action against anti-erosion activity but nothing concrete came out because of difference in the approach of going about it; after 3 years gap, in February 2010, secretary level meeting took place and 50 points have been identified for anti-erosion activities because of good understanding. Since January 2009 to February 2010, almost every month, technical level meetings have been taking place. You can understand that the political atmosphere plays an important role".[8]

Mr. Iyer has explained politicisation of water this way.

> Water disputes are inevitably political and cannot be removed from the domain of politics. However, they become 'politicised' in a negative sense when considerations of party politics and of impacts on elections bring in adversarial attitudes and come in the way of a constructive and cooperative approach to the management of shared natural resources.[9]

---

[6]Record of Discussions of the Thirty-Sixth Meeting of the Indo-Bangladesh Joints Rivers Commission Held at Dhaka from 19th to 21st September, 2005, <http://waterbeyondborders.net/files/Ind_ban_JRC_36_sep2005.pdf>, accessed June 17, 2012.

[7]Record of Discussions of the Thirty–Seventh Meeting of the Indo-Bangladesh Joint Rivers Commission held in New Delhi from 17 March to 20 March 2010, <http://waterbeyondborders.net/files/minutes_of_meeting/Ind_ban_JRC_37_mar2010.pdf>, accessed on June 17, 2012.

[8]Personal interview with the author on 15 June 2010 at New Delhi.

[9]Iyer (2008), p. 372.

In fact, it is the Indian states which have grievances against the continuation of the treaty. The Calcutta port has been complaining about low quantum of discharge of water from Farakka, especially during the dry season (January to May); their complaint is that discharge has been going down which influences sediment transport pattern in the river. Records reveal that the discharge from Farakka during the dry season fluctuates widely and, on most of days, it is much less than the desired quantum".[10] "The maximum quantity ever received was 30,000 cusecs".[11] This port caters to large client and needs of eastern and north-eastern states. If ship movement is affected due to siltation, it is going to cause damage to its users which include Indian Oil Corporation (importing crude through the dock for its both Haldia and Barauni refineries), Tamil Nadu Electricity Board (importing over three million tonnes of thermal coal through Haldia by way of coastal movement), SAIL and Tata Steel (importing large quantities of coking coal and exporting finished steel items), among many others in the hinterland covering vast areas in the UP, Bihar, Madhya Pradesh, West Bengal, North Eastern region and even the neighbouring countries of Nepal and Bhutan.[12]

Another provincial state in the east, Bihar government had strongly protested against the Ganges treaty even in 1996. The Bihar state government was so enraged with the Farakka treaty that it issued White paper containing annexed collection of letters and documents exchanged between officials of New Delhi and Patna before finalisation of the Treaty.

Bihar has complaint that Delhi signed the Farakka Agreement with Bangladesh without taking it into confidence. The White Paper of Bihar reads like this:

> On 13 December 1996, it is an unpleasant surprise to find that Bihar…(is) excluded from international and interstate talks on sharing the Ganges. From newspapers we learn that India and Bangladesh are going to have an agreement on water allocation. While Bihar has been kept in the dark about this, West Bengal has been provided the opportunity for full participation. The Chief Minister of West Bengal was instructed by the Centre to help finalise the treaty and even to prepare its draft. From Doordarshan TV we learn that the Bangladeshi Prime Minister has come to India to finalise the treaty, and that India has agreed to provide Bangladesh a minimum of 34, 500 cusecs of water.[13]

In fact, the United Front government had justified the exclusion of Bihar in the negotiation with Bangladesh. On being questioned about why only West Bengal government and its chief Minister was being consulted and not Bihar and UP's, the irrigation minister replied, "so far as the involvement of Chief Minister of West Bengal is concerned, the Calcutta port is going to be directly affected by the water flowing from Farakka. Hence it was in the interest of the state that the Chief

---

[10]Shantanu Sanyal, "Kolkatta Port: Government Support Vital," http://www.hindu.com/businessline/logistic/2001/11/19/stories/0919c05s.htm, accessed on 2 July 2012.

[11]Sau (1990), p. 1017.

[12]Sanyal, op. cit.

[13]Dipak Gyawali, "No one gets anything. Someone gets everything. Everyone gets something," http://www.himalmag.com/component/content/article/2178-No-one-gets-anything.-Someone-gets-everything.-Everyone-gets-something..html, accessed on 15 May 2010.

Minister of West Bengal should have intervened in this matter. He is a leader of the national level and it was duty of the government to entertain his suggestion". Though he accepted that Bihar and UP would be affected " it is true that during the lean season, these states would be urged upon and pressed to make efforts to ensure continuous flow of water".[14]

The Government of India has put certain restrictions on consumptive usage of water on Bihar as par the treaty provisions where it says India would try to make 'every effort' to protect flow at Farakka at any cost. That is why, Bihar has been demanding review of the Ganges Treaty. Bihar's present chief minister Mr. Nitish Kumar made a statement that "Farakka barrage was constructed and an accord was signed on sharing of Ganges water with Bangladesh and as the water was being regulated to Bangladesh, a unit of state-owned National Thermal Power Corporation (NTPC) shut down, affecting power supply to Bihar".[15] In response to persistent complaints from Bihar, the central government has changed the rules for water storage procedure which was denied to the state after 1996 commitment.

In 2009, in response to strong criticism from the state, the Ministry of Water Resources has changed the rules in water utilisation for thermal Projects of the Ganges Basin that suggests that the project authority can create the ponds for storing sufficient flood water during monsoon to meet their water requirements during the lean season. The project authorities may also use the ground water for meeting their requirements in accordance with the local ground water availability. "The changed guidelines further add that the water may be made available by the concerned states from their irrigation canals against the approved water allocation of such irrigation projects but states cannot claim any right for extra use of water on this account when the water share of the states in Ganges Basin is decided in the future".[16]

In fact, West Bengal also protested against the Ganges Treaty as extra water was being released to Bangladesh through Farakka. In 2012, the Chief Minister, Ms. Mamta Banerjee wrote a letter stating "the centre's apathy in repairing two damaged sluice gates of the Farakka barrage has led to release of Ganges water to Bangladesh much above what the neighbour is entitled to during dry season. This is affecting drinking water, shipping movement and electricity generation.[17] She

---

[14]The Irrigation Minister participating in the debate on Water Sharing Agreement Between India and Bangladesh in Lok Sabha, Lok Sabha Debates, vol. IX, no. 2, February 1997, New Delhi: Lok Sabha Secretariat, 1997, p. 20.

[15]"Review Ganges Water Treaty, Bihar CM asks Indian govt; says his state being deprived," http://www.thedailystar.net/newDesign/news-details.php?nid=202286, accessed on 14 September 2011.

[16]Notification regarding change of rule was given to the author by the Ministry of Water Resources in May 2012.

[17]"Water row brews, this time over Farakka," http://articles.timesofindia.indiatimes.com/2012-02-16/kolkata/31066042_1_water-row-farakka-barrage-sluice-gates, accessed on 20 February 2012.

further said, "we are suffering because of Farakka. We have compromised with this crisis for the sake of the greater interest of Bangladesh-India relationship…"[18] Thus Indian states are becoming restless about international commitments of the union government.

The year 2016 saw very low flow in the Ganges in some 10-day period of March. As a result, the NTPC's plant near the barrage had to shut operations from March 10, 2016. The inflow at the Farakka barrage was nearly half, compared with the quantum of water available in the last 2 years. The newspaper quoted the data obtained from the Superintending engineer at the Farakka barrage, who shared that "on March 29, the inflow discharge observed in the Ganges by the India-Bangladesh joint observation team was 50, 710 cusecs, while in comparison the discharge on 29/03/14 and 29/03/2015/ were 91,001cusecs and 83,807 cusecs respectively".[19] Soon, these complaints might turn into major protests when full impact of the climate change becomes visible on the Ganges.

5.3 Climate Change and Impact on Ganges River Water Treaty

The glaciers of the Himalayan mountain ranges are at the centre of Asia's water supply because they provide a life-support system to millions of people living in South Asia. They release water slowly into the vast network of rivers of the Indus, Ganges and Brahmaputra basins. But over the years the general trend in the area suggests that glaciers are thinning at an average rate of approximately 1 m per year. Annual glacial melt water from the Himalayas in spring and summer, which is replenished during winter snowfall, accounts for as much as 70 % of the summer flow in the Ganges River and 50–60 % of the flow in Asia's other major river systems. Recent measurements show that there is a regression of approximately 30 days in the maximum spring flow period and an increase of 30–38 % in the glacial runoff. This implies that in the span of years, the Himalayan rivers that depend on glacial melting—such as the Ganges, Indus, Brahmaputra and Mekong River systems—will lose perennial nature and become seasonal rivers when the glaciers melt. As the shortage arrives in approximately 20–30 years, it will be quite abrupt, as the flow will be dramatically reduced in the dry season.[20] Today's creation of robust structure will provide smooth handling when the calamity strikes.

Moreover, the Ganges moves through a densely populated region, a population of almost 500 million live in the basin which is more than one third of India's population. Other regions of South Asian countries where the Ganges flows, also

[18]"Bangladesh getting extra water through Farakka Barrage," http://bdnews24.com/details.php?id=218849&cid=2, accessed on 10 July 2012.

[19]Venkat (2016), p 1.

[20]Pachauri (2009), p. 33.

have a heavy burden of population. Bangladesh has 37 million, Nepal has 23 million people. Of the total drainage area of 1,087, 300 km^2 of Ganges, about 860,000 km^2 fall within India, 46,300 km^2 in Bangladesh and 1,47,480 km^2 in Nepal. This suggests that the Ganges is very important river from drainage as well as number of people being supported by the river for India, Bangladesh and Nepal.

In contrast to traditional assumption that the past hydrological experience provides a good guide to future conditions, the phenomenon of climate change may alter the reliability of current water management systems and water-related infrastructure. By and large, dry-season flow in the major Himalayan rivers in a given year results from the monsoon rainfall of the previous year. If there is any climatic change in the mountain hydrological regimes, it is likely to alter these resources, and severely affects lower riparian Bangladesh that depends on this water resource. From the decadal trends in the rainfall variability in Bangladesh, it has been observed that significant deviation of monthly rainfall from one decade to the other has occurred, and the variations of total seasonal rainfall, the timing of onset, peak and recession have changed considerably. As a result, the scarcity of pure drinking water is gradually increasing at a dramatic pace in both urban and rural areas in Bangladesh.[21]

Notwithstanding emerging scenarios, practice of farming has been changing both in India and Bangladesh. In Bangladesh, about 84 % of the total population is engaged in the rural sector; 61 % of the total land area is arable and 4/5th of the arable area is under paddy cultivation, thus, water for agriculture is vital. Farming is increasingly becoming dependent on irrigated agricultural practices. Aus, a pre-monsoon variety, is being replaced by boro-(irrigated) dry season rice. Aus is no longer an important crop as boro has become the crucial crop after the introduction of High Yielding Varieties. Now Boro contributes to nearly 60 % of the total rice production. This implies that the demand for water is further expected to rise from irrigation expansion. According to World Bank 2006 record, 7.6 million hectare of the total cultivable land of 9.03 million ha used in agriculture are suitable for irrigation and about 4.5 million ha are irrigated.[22] The private sector provides about 90 % of this irrigation which comes mainly from groundwater. It is expected that the irrigated area would reach 6.9 million hectare by 2020.[23]

It is anticipated that climate change will affect Bangladesh through submergence, salinity intrusion and destabilisation of rivers. Restoration of full flow of Bangladeshi rivers is a must to resist salinity intrusion. Rivers have brought in about 2 billion tons of sediment across the time, which has raised its surface by about 2 mm each year. Continuation of this sedimentation process is a must for

[21]Md. Rashed Chowdhury, "Water: The New Oil," http://www.thedailystar.net/newDesign/print_news.php?nid=213847, accessed on 13 December, 2011.

[22]Justine Treadwell and Ali Shafqat Akanda, "Contributing Factors in the Ongoing Water Conflict Between Bangladesh and India," https://wikis.uit.tufts.edu/confluence/display/aquapedia/Contributing+Factors+in+the+Ongoing+Water+Conflict+Between+Bangladesh+and+India, 21 October 2011.

[23]World Bank (1998).

Bangladesh to withstand the submergence effect. But with reduction of river flow, sediment flow will also reduce. Reduced flow will raise the potential of salinity intrusion. Water Resource Minister shared in the Parliament that about one-third of the country's rivers are almost dead. The minister further added that one of 310 rivers of the country was completely dead while 97 were drying up. The country's rivers are drying and losing navigability for lack of dredging; 30,000 km of waterways in the country were reduced to a mere 3500 km.[24]

Another important problem being faced by Bangladesh is river erosion and filling up of river bodies. As climate change will intensify the process of river erosion, this has become an urgent issue for India and Bangladesh. In the Joint Rivers Commission meeting in 2006, Bangladesh has raised the issue of having lost 900 acres of land due to erosion of the border rivers of Ichhamati and Kalindi in Satkhira. Sometimes new land emerges along the rivers and this leads to controversy over their ownership. Border forces from both sides claim them which create acrimonious situations. As for an illustration, Indo-Bangladesh dispute over New Moore Island-a small uninhabited offshore sandbar landform in the Bay of Bengal, off the coast of the Ganges-Brahmaputra Delta region. Both India and Bangladesh claimed sovereignty over it because of supposed existence of oil and natural gas in the region.

On the basis of the 2011 population census, India's per capita availability of water has fallen below the global threshold, signalling that the country would have to address conservation needs more seriously in consideration of a growing population and an expanding economy. India's per capita availability has been pegged at 1,545 m^3 a year, including non-personal consumption such as irrigation according to an estimate of the water resources ministry which is below the international threshold of 1,700 m^3 a year. According to UN-adopted Falkenmark water Stress indicator, the most widely cited measure of water scarcity, national per capita availability indicates water stress conditions.[25]

Another climate change-induced concern is the degradation of the Sundarban mangrove forest—one of the largest such forests in the world. Of the 9,600 m^2 of Sundarbans, 40 % lie in India and the rest in Bangladesh. The Sundarbans are critical for the protection of low-lying areas because they provide stability in the face of high tides, holding the trees and also absorbing salt. As India and Bangladesh are already sharing data on flood forecasting, new stations can be set up to expedite the transfer of data on other aspects of environmental degradation.

That is why, Bangladesh keeps a hawkish eye on news related to water development on the Ganges. The news which attracts Bangladesh attention and reaction is the publicised planning of inter-basin transfer of water in India, this is popularly known as inter-linking of river projects.

---

[24]http://www.thedailystar.net/newDesign/news-details.php?nid=222372, accessed on 14 February 2012.

[25]http://www.hindustantimes.com/India-news/NewDelhi/India-facing-water-crisis/Article1-779728.aspx, accessed on 10 December 2011.

5.4 Interlinking of River Projects and Bangladesh's Concerns

The interlinking of river project is an old idea which keeps getting revived on regular intervals. About 71 % of the available water resources of India is localised in 36 % of the geographical areas of the country, primarily in the Ganges-Brahmaputra Meghna basin and all the West flowing rivers from the Western Ghats. The main idea is to transfer water from the Ganges-Brahmaputra river system to the less water endowed areas in southern and western parts of India by linking canals. The National Water Development Agency (NWDA) was given the task of preparing detailed study which proposed 30 links in the Himalayan and Peninsular components.[26] This project was revived by the National Democratic Alliance government in 2002.

The Supreme Court judgement on February 27, 2012 directed the Indian government to implement the interlinking of river projects in a time bound manner and further appointed a high-powered committee for its planning and implementation. Speaking on the interlinking of river projects, Deputy Chairman of the Planning Commission, Montek Singh Ahluwalia said that "effective exploitation of capacity of these rivers is critically dependent on the international agreements." He further added "it is not a decision India can take alone".[27]

Groups of people have filed a review petition to the Supreme Court. But implementation of the river-linking project does not seem possible at the moment or in near future because almost all intra-state rivers are involved in conflict between different provinces on sharing issue, most celebrated ones among them are Cauvery river dispute between Tamil Nadu and Karnataka, Yamuna link canal between Punjab, Haryana and Delhi which have been hanging unresolved for ages. In this case at least Bangladesh can be relaxed and does not need to become concerned because it is a battle in India which has to be fought inside before it spills over across border.

5.5 Emerging Role of Provinces in Indian Federation

Division of power is between the union and provincial state governments. Indian Political system has evolved from a 'single party dominant political system' to an era of emergence of multiparty system now. There was a time when India was known as "…a Unitary state with subsidiary federal features rather than a Federal

---

[26]For details on the issue see, (Iyer 2003), Alagh et al. (2006).

[27]"Ahluwalia cautions on inter-linking of rivers," http://www.thehindu.com/news/national/article3300168.ece, accessed on 11 April 2012.

State with subsidiary unitary features".[28] It is moving towards a centre with strong federal provincial states which are very competitive about protecting its interests.

With the advent of coalition politics since 1989, regionalisation of Indian politics has taken place. This has led to a formation of coalition governments both at the centre as well as states. Consequently, commanding position of the central government has weakened even in foreign affairs where the union government has exclusive power.

There is a division of power between the centre and the states—union list, state list and concurrent list. Parliament has exclusive power to legislate on 18 items enumerated in Seventh Schedule of the Union list. Foreign affairs come under this. It enumerates all matters which bring the Union into relation with any foreign country, entering into treaties and agreements with foreign countries and implementation of treaties, agreements and conventions with foreign countries.

There are other references in the constitution which give the Federal Government full power to match their responsibility in the matter of foreign affairs. According to Article 253, Parliament has power to make any law for the whole or any part of the territory of India for implementing any treaty, agreement or convention with any other country or countries or any decision made at any international conference, association or other body. Article 73 also reads that the Union has full executive powers for the implementation of its laws, the treaties and agreements it might make. Another article which is related to foreign affairs is article 257 which prevents the States from obstructing the Union's administration of its laws, it is laid down that the executive power of the states is to be so exercised as not to impede or prejudice the exercise of the executive power of the Union. Further, the executive power of the Union also extends to the giving of such directions to a state as may appear to be necessary for that purpose.

Even though the central government has been empowered by the constitution to conduct foreign affairs relationship freely; the extent of the co-operation required from state Governments varies according to the nature of the subject—whether it is exclusively under the jurisdiction of the Union Government, a concurrent subject, or exclusively under the jurisdiction of the states. The regulation of the inter-state river is vested in the Union government but nothing is written on sharing of water with neighbouring countries even though the issue of Indus was raised in 1948.[29]

Also within India, so many intra-state river water disputes are going on. Solution to these disputes is not looking feasible. In fact, the responsibility for water issues is so fragmented within the federal government that more than a dozen departments or ministries deal with various segments of water resources. Getting agreement on any water issue involving so many ministries and department is itself a huge task. This kind of arrangement between the centre and states create confusion even for neighbouring countries.

---

[28]Who are quoted in Durga Das Basu, "Nature of the Federal System," Introduction to the Constitution of India, New Delhi: Prentice-Hall of India, 1997, p. 59.

[29]Swain (2010), pp. 27-34.

In fact, the change of relationship between the centre and state level leadership can be brought here with an illustration albeit in another case. In order to remove the tension and border conflicts, the Prime Ministers of India and Pakistan had entered into an agreement called the Indo-Pakistan Agreement for an equal division of Berubari Union between India and Pakistan on September 10, 1958. But the decision of union government created lots of hues and cry in the West Bengal. In a statement in the Lok Sabha on September 12, 1958, Nehru said: "we have come to an agreement and after all we have to honour it. If I have made any mistake you can censure me but I have to honour it so far as I am concerned"[30]

Speaking during the debate on Berubari in the State Assembly, the chief Minister of West Bengal, Dr. B.C. Roy made it clear that so far as he, his party and his Government were concerned, they would stand by the Prime Minister's pledge to honour the agreement reached with Pakistan.

> Right or wrong, the Prime Minister (of India) has come to an agreement with the Pakistan Prime Minister. I do not want it to be said in the world that our country did not honour its words. Mr. Nehru is, after all, the Prime Minister of the country and I cannot lower the prestige of the Prime Minister.[31]

The present situation is in complete contrast of that time. The decision of India not to sign an agreement on Teesta river during Indian Prime Minister's visit to Dhaka in September 2011 entails changed dynamics of domestic politics of West Bengal. In fact, previously the visit of Indian Prime Minister to Bangladesh was postponed for provincial election of West Bengal. Once election was over, it was assumed that any decision on Teesta river could be taken without too much politicisation. United Progressive Alliance (UPA) partner Trinamool Congress formed the provincial government along with Indian National Congress as coalition partner in West Bengal in 2011. Both India and Bangladesh governments had been working consistently on Teesta agreement since Bangladesh Prime Minister Sheikh Hasina's visit to India in January 2010. At least four rounds of water secretary and technical level meetings took place to fine-tune the agreement.[32] It was agreed between the two sides that when Indian Prime Minister Manmohan Singh made a visit to Dhaka in September 2011, an interim agreement on Teesta would be announced. But West Bengal Chief Minister decided not to accompany Prime Minister and raised serious objection about framework of ensuing Teesta interim agreement.[33] The Chief Minister Ms. Bannerjee argued that she was not sufficiently consulted before committing certain amount of water to Bangladesh which would

---

[30]Quoted in Appadorai (1981), 192.

[31]quoted in ibid., p. 195.

[32]"Teesta draft fine tuned," http://www.thedailystar.net/newDesign/news-details.php?nid=188968, accessed on 15 December 2011.

[33]"Mamata Banerjee refuses to go to Dhaka with PM," http://articles.timesofindia.indiatimes.com/2011-09-05/india/30115583_1_teesta-water-bangladesh-visit-mamata-banerjee, accessed on 14 November 2011.

harm interests of farmers of North-Bengal.[34] Thus the Teesta Agreement was not signed and this led to huge embarrassment to Indian government. Thus Indian political system has come a long way from the political atmosphere prevailing in 50s and 60s where reputation of the Prime Minister would be a good enough reason for accepting any decision of the central government by the state even though it considered that its interests are being harmed by the particular decision.

As discussed above, the Indian provincial states are not interested to lose or compromise its own source of river water. Over the past year the Union Minister of Water Resources has remarked, "I really am not Minister of Water Resources but Minister of Water Conflicts", and the Union Finance Minister has noted a "growing set of small civil wars" over water at all levels in Indian society mostly between users of drinking water, irrigation and industry.[35] Thus India has serious internal as well as external river water disputes, some innovative and alternative options are required to deal with future challenges.

5.5.1 Exploration of Alternatives

Now both India and Bangladesh are becoming conscious of climate change phenomenon and trying to address the issue in their own way. It is learnt that Bangladesh is planning to construct the Ganges Barrage. This idea was first raised by Pakistan in 1963 but India had opposed the barrage arguing that large areas of Indian Territory in the state of West Bengal would be submerged as a result of backwater effect. After conclusion of the Ganges Treaty and Bangladesh's agreement to change especially its location India has agreed to the construction of the barrage. This was confirmed in the Joint Communiqué of the Joint Rivers Commission in its 32nd meeting which was held on July 18–20, 1997. The Joint Communique stated:

> The Commission welcomed the proposal of Bangladesh to implement the Ganges Barrage Project. India indicated its intention to consider providing technical assistance through Water and Power Consultancy Limited (WAPCOS), a Government of India Undertaking, which has the requisite expertise in this regards.[36]

Now the location has been shifted to Pangsha, 90 miles west of Dhaka, and not the Hardinge Bridge as originally planned. The government plans to begin construction of the Ganges Barrage in Rajbari district in December 2012 to preserve the river water during monsoon and release it during lean period. The barrage will

[34]The election election 2011 assembly suggests that the performance of Ms. Party was not good in North Bengal in comparison to other parts of state so she refused to be part of unpopular decision because Teesta irrigates North Bengal.

[35]"Water Ministry seeks World Bank Funding for reforms," *The Hindu*, 13 January 2005.

[36]Quoted in "The Ganges River," (Salman 2002), 187.

facilitate irrigation of about 19 lakh hectares of arable land in greater Kushtia, Faridpur, Jessore, Khulna, Barisal, Pabna and Rajshahi districts.[37] The study suggests that the flows in all tributaries and other rivers in the southwest region would be augmented with the Ganges barrage so that the natural environment, like fisheries, groundwater, forestry, human health and navigation could be restored through the supply of upland water flow and a reduction in salinity.

In India, the National Water Mission as part of National Action Plan on Climate Change takes into account the effects of climate change. It underlines the need to priorities documentation and modeling of surface and groundwater resources in each river basin to benefit users of each basin. Also as part of new approach it is suggested that a comprehensive map of the country's aquifers with their storage and transmission characteristics at a watershed scale would be prepared. This would help crop water budgeting. If each region or river basin is planned according to all the parameters which have been enshrined in the Planning Commission document, the planning and execution of water related development projects will be in congruence with tolerable capacity of the surface region, this is going to generate positive spillover effect to other neighbours. Then India does not need to resort to overexploit surface and sub-surface water. On the one hand, this leads to conveyance of good flow of water to Farakka barrage and on the other hand, India can also assist Bangladesh in preparation of sustainable water projects if neighbour asks for. So Bangladesh would be able to prepare water development as well as cropping pattern according to its climatic and geographic features, this is the only way forward considering impending climatic change effect. There has been some positive development between neighbours about river water cooperation when India decided to offer Bangladesh an equity stake in the Tipaimukh Dam in Manipur. In 2011, as part of India's credit of one billion for infrastructure development to Bangladesh, dredging of Ichamati river has begun. The Bangladesh government has also put dredging of rivers in fast-forward mode intending to cover 300 rivers by 2025.

At the regional level in the 16th SAARC Summit in Thimpu, the leaders of eight SAARC countries agreed to establish an inter-governmental expert group on climate change for developing and monitoring regional policy implementation. They also decided to plant ten million trees in the SAARC countries in 5 years as part of a regional attempt to afforestation and reforestation.

The unusual seasonal variations of water is one of the main characteristics of the Ganges river. The Ganges has nine tributaries joining it from Nepal, three of which—Karnali, Gandaki and Kosi—arise in Tibet. These three rivers Karnali, Gandaki and Kosi supply more than 70 % of the dry season flow. It is known that China has already constructed many dams and some more are proposed which would badly affect the flow of these rivers. Though China claims that these dams would be used for power generation and would not affect the flow to lower riparian countries like India and Bangladesh. Here India and Bangladesh would have to make concerted effort to protect their interests. But as of now, India has been expressing its concerns alone.

[37]"Ganges barrage work from 2012."

The Ganges river basin is one of the most fertile and densely populated in the world. The river flows through 29 cities with population over 100, 000, 23 cities with population between 50,000 and 100,000 and about 48 towns. As already mentioned, total population which resides in the basin is more than 500 million. The river is lifeline for millions of people who depend on the various socio-cultural and ecosystem services offered by the river and its vast basin.

As a result, some sort of realisation is dawning in the government that indiscreet dam building is going to invite environmental disaster. In 2014, the central government has accepted to the Supreme Court that the 2013 Uttarakhand floods and landslides were linked with the development of hydropower projects. This led to cancellation of 24 out of 39 proposed projects in Uttarakhand after noting that they significantly impacted biodiversity in two sub-basins of the Alakhanda and Bhagirathi rivers.[38]

Another area of consensus is emerging between both India and Bangladesh is that they shall leave 20 % of the river flow untouched to preserve circulation. This recognition is a significant departure from their previous position that water that is left to flow into the sea should be considered as 'waste'. Recognition of the right of survival of the river will set a precedent for negotiations on other 53 rivers that India and Bangladesh share. The emerging understanding is that rivers should not be exploited to an extent that their survival would be at stake.[39] Thus maintaining ecological health of a river has become important subject.

In 2014, Government of India started an integrated Ganges development project titled, 'Namami Gange' for the rejuvenation of Ganges river. The program is focused on maintaining continuous flow as well as pollution abetment interventions namely interception, diversion and treatment of wastewater flowing through the open drains and other means.

5.6 Conclusion

Thus, above descriptions draw a mixed picture about future of the Ganges river issue. Future of the Ganges treaty in 2026 will depend not only on river water but also on socio-economic and political transformation of the region. How people are becoming conscious of the environment and follow agro practices which are in synergy with natural climatic and soil capacity. This kind of attitude would lead to sustainable planning. On the political side, one of the important aspects which emerges is that whenever the ruling regimes in New Delhi and Dhaka are on friendly terms and either side is not capitalising the water issue for its domestic political ends, prospects of finding a mutually acceptable agreement becomes brighter.

---

[38]Utkarsh Anand, Government Admits: Hydro Projects did affect Uttarakhand floods, Indian Express, December 9, 2014.

[39]Pandey (2011).

It is important to conclude here that the 1996 treaty does not offer an ideal solution. It has been criticised on both technical and political grounds but the very fact that it became possible to negotiate and sign a long-term treaty in the first place and then, it has been working uninterrupted for almost two decades is a huge breakthrough.

References

Alagh KY et al (2006) Interlinking of rivers in India: overview and Ken-Betwa link, Academic Foundation, New Delhi

Appadorai A (1981) Federalism. In: The domestic roots of India's foreign policy 1947–1972. Oxford University Press, Delhi, p 192

Elhance, AP (1999) Hydropolitics in the third world conflict and cooperation in international river basins. US Institute of Peace, Washington DC

Iyer R (2003) Water perspectives, issues, concerns. Sage Publications, New Delhi

Iyer R (2008) Transboundary water conflicts. In: Joy KJ, Biksham G, Suhas P, Vinod GSV (eds) Water conflicts in india a million revolts in the making. Routledge, New Delhi, p 372

Joy KJ, Gujja Biksham, Paranjape S, Vinod G, Shruti V (eds) (2008) Water conflicts in India: a million revolts in the making. Routledge, New Delhi

Pachauri RK (2009) Climate change and water, Asia's next challenge: securing the region's water future: a report by the leadership group on water security. Asia Society, Mumbai, Asia, p 33

Salman SMA, Kishor U (2002) The Kosi river. In: Conflict and cooperation on South Asia's international rivers a legal perspective. Kluwer Law International, The Hague

Swain A (2010) Environment and conflict in South Asia: water-sharing between Bangladesh and India. South Asian J 28:27–34

Swain A (2012) Global climate change and challenges for international river agreements. Int J Sustain Soc 4(1/2):72–87

Venkat V (2016) Is the mighty ganges drying up? 6(14):1 (City edition, The Hindu)

Website References

Anand, Utkarsh Government Admits: Hydro Projects did affect Uttarakhand floods, Indian Express, December 9, 2014

Bangladesh News 24. (2012). Bangladesh getting extra water through Farakka Barrage. http://bdnews24.com/details.php?id=218849&cid=2. Accessed 10 July 2012

Chowdhury, Md. Rashed (13 December 2011). Water: The New Oil. The Daily Star. http://www.thedailystar.net/newDesign/print_news.php?nid=213847. Accessed 13 December 2011

Ganges barrage work from (2012), The Daily Star. http://www.thedailystar.net/newDesign/news-details.php?nid=157349. Accessed 4 July 2012

Gyawali D (1999) No one gets anything. Someone gets everything. Everyone gets something. http://old.himalmag.com/component/content/article/2178-No-one-gets-anything.-Someone-gets-everything.-Everyone-gets-something..html. Accessed 15 May 2010

Treadwell J, Akanda AS (2011), Contributing Factors in the Ongoing Water Conflict Between Bangladesh and India. https://wikis.uit.tufts.edu/confluence/display/aquapedia/Contributing+Factors+in+the+Ongoing+Water+Conflict+Between+Bangladesh+and+India. Accessed 21 October 2011

Mamata Banerjee refuses to go to Dhaka with PM. http://articles.timesofindia.indiatimes.com/2011-09-05/india/30115583_1_teesta-water-bangladesh-visit-mamata-banerjee. Accessed 14 November 2011

Pandey P (2011). India-Bangladesh Relations: Need for New Approach, No. 127/2011. http://www.rsis.edu.sg/publications/Perspective/RSIS1272011.pdf. Accessed 5 September 2011

Review Ganges Water Treaty, Bihar CM asks Indian govt; says his state being deprived. http://www.thedailystar.net/newDesign/news-details.php?nid=202286. Accessed 14 September 2011

Sau SN (1990) The economics of Calcutta-Haldia Port complex. Eco Polit Weekly, p 1017

Sanyal S (2001) Kolkatta Port: Government Support Vital. http://www.hindu.com/businessline/logistic/2001/11/19/stories/0919c05s.htm. Accessed 2 July 2012

Teesta draft fine tuned. http://www.thedailystar.net/newDesign/news-details.php?nid=188968. Accessed 15 December 2011

The Hindu (10 April 2012). Ahluwalia cautions on inter-linking of rivers. http://www.thehindu.com/news/national/article3300168.ece. Accessed 11 April 2012

Water Ministry seeks World Bank Funding for reforms, *The Hindu*, 13 January 2005

Water row brews, this time over Farakka. http://articles.timesofindia.indiatimes.com/2012-02-16/kolkata/31066042_1_water-row-farakka-barrage-sluice-gates. Accessed 20 February 2012

World Bank (1998) Bangladesh 2020: a long-run perspective study, The World Bank, Bangladesh

Chapter 6
Analytical Observations and Conclusion

Democratic interests in conflict and power struggles often drive water policy in transboundary basins sometimes more so than contending interests between countries.[1] The analysis of the Ganges issue underlines the centrality of domestic politics influencing the trajectory and outcome of the negotiation. Thus the river has not only been about hydropolitics but has turned into hydroelectoral political issue especially in Bangladesh where river water controversy is used as an instrument to beat each other for winning over constituencies for votes.

It can be observed after the examination of a quarter of a century negotiation before 1996 and post treaty period that in South Asia, people are conscious of power, prestige and protocol. In the run up to final negotiation, West Bengal and its chief minister were made important stakeholder. Seconding of the treaty by Jyoti Basu buttressed the effort of the Union government to get wider acceptance in India's domestic Politics. This example can be emulated for resolution of other bilateral river disputes including the Teesta.

As discussed earlier, river water issue is not independent of overall bilateral relationship. This can be demonstrated with an illustration. Inspite of repeated demands from Pakistan for ministerial talk to discuss farakka, India did not accept this. An interview can be cited here where Mrs. Gandhi confessed that political level talk could not be held with Pakistan because overall relationship was not conducive for this. "… Talks can only be held if there is an atmosphere of confidence and amity. That was a period when there was acute distrust, and no respect, therefore there was no agreement even on much lesser issues".[2]

Another important observation can be made is that previously Pakistan and later Bangladesh tried to internationalise the Ganges issue. But there is a limitation of multilateral agency's role in a bilateral river dispute. It is worth mentioning here

[1]Importance of Transboundary Waters (2014), P. 2.
[2]Crow et al. (1995).

P. Pandey, *India Bangladesh Domestic Politics*,
DOI 10.1007/978-981-10-2371-2_6

that Bangladesh brought the farakka issue to the General Assembly broadly under the provisions of Article 14 of the UN Charter. Article 14 of the UN charter reads like this:

> the General Assembly may recommend measures for the peaceful adjustment of any situation, regardless of origin, which it deems likely to impair the general welfare or friendly relations among nations, including situations resulting from a violation of the provisions of the present Charter setting forth the Purposes and Principles of the United Nations.[3]

So, there is not a single provision in the UN which suggests involvement of the multilateral body in the water dispute between two parties. Any party can bring the issue in front of the world body as derivative or extension of the prevailing article of the Charter.

The 1996 treaty has to face the climate change phenomenon and its various impacts. Visible changes of Climate change like low availability of water and other impacts have already discussed in the fifth chapter; the invisible impact would be in the form of acrimonious relationship between India and Bangladesh. India would have to face hostile perception battle even though progress might be happening on other fronts. As for an illustration, during 2011 Indian Prime Minister Manmohan Singh visit, India signed Land boundary agreement but this important breakthrough was swallowed by the Teesta news when Agreement on the river was not signed. The visit did not elicit goodwill for the country as was expected. Since India and Bangladesh share another 53 rivers, cooperation on all these rivers would be negatively impacted.

Another important aspect which comes in the discussion of river water dispute between India and Bangladesh is that there should be transparency of data between the two countries. Transparency is important but a complex issue. One of the reasons for not having agreement between the BNP and the Congress government during the early 1990s was because quantity of water being demanded by the Bangladesh government was not available in the Ganges. The Zia government has got 34,500 cusecs, Bangladesh was demanding at least that much water but this much water was not available in Farakka. Since the quantity agreed by the previous regime known, it was politically suicidal for other government to go for a negotiation which would give less than previously known quantity. Compromise is possible if the political leadership is able to calibrate its policy and convince the domestic constituency about the changed reality. But in the prevailing political culture, expecting this from political leadership is audacious. In 1996, political engineering was administered in the statistical data of water by taking an average of large range (1949–1988), so that figure looks great and gratifying to people. It is believed that agreement over 35,000 cusecs was more of a data engineering than reality that gave a bigger average. It might be difficult to extend the treaty with the same quantity of water because of climate change in 2026. It can be hoped that functional democracy prevailing in both countries will graduate into mature and substantive democracy where people would not be largely guided by emotions but

[3]http://www.un.org/en/documents/charter/chapter4.html, accessed on 28 June 2012.

appreciate the compromises which governments would have to undergo because of changed situations when the treaty term gets over.

The 1996 treaty is premised on distributive bargaining approach where river water is assumed to be fixed and will remain so. That is why fixed schedule has been prepared for thirty long years. But the reality is different. Water level keeps fluctuating from 1 year to another and from one 10-day lean period to another. It is common complaint among Bangladeshis is that India has not been true to schedule. In some lean period, Bangladesh gets more water than scheduled, other days less as chapter five discussion suggests. India responds that the schedule is indicative, if the average water level is available, then only schedule can be maintained.

If India and Bangladesh had adopted integrative bargaining, reality would have been different. Integrative bargaining would have addressed augmentation issue where both countries have had different preferences. This would have intensified exploration of alternative methods for augmentation more intensely in the less heated moment of post treaty period. Cooling period is always good time to discuss differences. Although the treaty recognises the need for cooperation by the two governments "in finding a solution to the long-term problem of augmenting the flows of the Ganges/Ganges during the dry season", it has not laid down any terms of reference or time frame for a joint study, nor has it created or entrusted any committee with the responsibility of carrying out this study. Detailed reading of the JRC meetings suggest that it has not ever been discussed how to go about augmentation. There has not been any study which suggests augmentation needs big interventions like construction of dam in Nepal or small remedial measures would help the situation.

Integrated planning also requires a radical change in the mindset from both sides of the border where water is understood in acquisitive sense. This kind of understanding has grown on prejudices and idiosyncrasies of both societies. The integrated approach is epitome of the culmination of trust and understanding between two neighbours which is not the case at the moment.

The preamble of the Ganges Treaty affirms the desire of both parties that waters from international rivers flowing in the territories of India and Bangladesh be shared between them by mutual agreement; water resources of the region should be put to the most beneficial use regarding flood management, irrigation use, river basin development, and the development of hydropower, but nothing concrete has been done. Though flood forecasting cooperation has been deepened, cooperation in other areas needs to be strengthened. Since livelihood of the Ganges region especially in South-Western part of Bangladesh is heavily dependent on the river, anything less than cooperation would generate large inflow of economic migrants, mostly towards India. Economic migrants will become a political tool in electoral politics of India in regions where social harmony is always precarious.

In light of il/legalities associated with immigration, large flow of immigrants will further complicate domestic as well as bilateral political relationship between India and Bangladesh. For example, political contestations along hyper-nationalism will polarise public opinion where even track-II efforts become difficult to organise. Another important related aspect is that India is already present in the domestic politics of Bangladesh. Distress in the country arising out of non-cooperation of India will be used between the Bangladesh Nationalist Party and the Awami League with allegations ranging from “compromises to national interest to selling the country’s interests”. This is not good for healthy neighbourly relationship.

The detailed examination of handling of water deficit issue in 1997 and after should provide the insights about improvements to be made to address the situation better. As a result of the low flow in 1997, an expert level meeting of the Indo-Bangladesh Joint Commission recommended the formation of a scientific committee to study the causes of the low flow of the Ganges river during the critical period of the dry season. But it is not known whether committee was formed or not, if formed how effective suggestions were, whether suggestions have been executed or not to improve the implementation of the provisions of the treaty?

What emerged out of close interactions with senior and retired bureaucrats, former ministers, activists, academicians, media personnel and members of advocacy groups in Bangladesh during track II meetings is that the successful negotiation of the issue depends not only on diplomatic deliberations but also on mode, behaviour, manner and protocol accorded to the visitor. The common perception among Bangladeshi diplomats have been that Indian foreign policy establishment did not show the same warmth and respect to them in the earlier days as they did to the western diplomats. This also, most probably, hardened the attitude of opposite groups in the negotiation. Thus, sentiments, emotions and perception hold as much importance for successful outcome as much as actual gain from the particular negotiation.[4]

It is generally assumed that the issue of Ganges river water sharing between India and Bangladesh has been resolved with the signing of Ganges Treaty in 1996, this can only be considered a stop gap arrangement in the long run of the river water dispute, only 2026 will decide how all the stakeholders, the government of India, government of Bangladesh, government the West Bengal and all concerned political parties are going to deal with the situation because by that time climate change impact will be far more visible in the Ganges than at the moment. Future of the Ganges river water treaty will also depend on availability of the river water but also on socio-economic and political transformation of the region.

Recently in India, some initiatives have been taken in the river water management which would likely to have strong bearing on the Ganges. For any irrigation or power project, it has been mandated that the design should be such in which at least one stream of the river should keep flowing. This ecological flow is important for keeping the river alive and will help better conveyance of water to the end.[5]

Another positive development seems to be new water policy of India. The draft of new water policy underlines the establishment of an institution to deal with

---

[4]The author has been part of track II interactions from 2009 to 2011.

[5]“Don’t design a dam on the river, make it on the side of the river, so that monsoon water can be captured and the river bed stays unharmed.” The minister was quoted in The Sunday Express, New Delhi, April 3, 2016, p. 10. vol. XV NO. 3.

interstate water disputes. The new policy suggests that the upper basin state shall adopt a cautious and minimalist approach to major interventions in the interstate rivers. This stresses that none of the states in a basin ‘owns the river’.[6] Though this is for domestic purpose of India but if implemented, this would definitely lead to better flow in the Ganges.

Thus, new institutional regime is required to deal with emerging challenges on a sustainable basis. If both countries keep postponing deliberations on issues of water quality, flood management and other impacts of climate change which might not seem pressing at the moment, will acquire fierce forms in future. For discussion on all these issues, various stakeholders need to be involved to build consensus. The river water relationship between riparian has to go beyond government and Ministry of External Affairs.

In the end, it can be concluded that it is important to say that it is time to institute a new structure or energise the existing structure which swing into action when there is low flow or in any other emergency demand. Obviously overall positive atmosphere thwarts the escalation of tension and accommodate others’ needs. But this cannot be the substitute for robust mechanism to sort out the irritants if any arises especially in the context of unpredictable climate conditions.

References

Crow B et al (1995) Sharing the Ganges the politics and technology of river development. Sage, New Delhi

Draft Water bill proposes ‘water for life’ for all, The Indian Express, June 3, 2016, vol LXXXIV, no 176, p 10

http://www.un.org/en/documents/charter/chapter4.html. Accessed 28 June 2012

Importance of Transboundary Waters, Benjamin (2014) The Rise of Hydro-Diplomacy Strengthening foreign policy for Transboundary Waters, Adelphi. Berlin, p 2

[6]Draft Water bill proposes ‘water for life’ for all, The Indian Express, June 3, 2016, p. 10.vol. LXXXIV NO. 176.

Index

A
Abbas, B.M., 7
Ali Bhutto, Zulfikar, 6
Augmentation, 8
Azam, Ghulam, 10

B
Bahini, Mukti, 10
Bangladesh, 1
Bangladesh high commission, 13
Bangladesh nationalist party (BNP), 10
Bangladesh white paper, 11
Barak, 12
Basu, Jyoti, 24
Bay of Bengal, 2
Beneficial Bilaterialism, 77
Bhagirathi-Hooghly, 2
Bilateral relationship, 1
Brahmaputra, 2
British India, 4
British, The, 3

C
Calcutta city, 2
Climate change, 118
Congress Manifesto, 13

D
Dhaka, 1
Desai, Morarji, 11
Dihang, 12
Domestic political, 1
Downstream, 2
Dry season flows, 12

E
East Bengal, 6
East Pakistan, 3, 5

F
Farakka Accord, 85
Farakka barrage, 1
Farakka day, 16

G
Gandhi, Indira, 5
Gandhi, Rajiv, 17
Ganges, The, 1
Ganges-Kobadak project, 3
Ganges-Brahmaputra–Meghna, 12
Gangotri, 2
General assembly, 14
Guaranteed supplies, 5

H
Hardinge Bridge, 27
Headwater supplies, 3
Himalayan glaciers, 107
Hooghly river, 1

I
Indian government, The, 3
Indian Prime Minister, 4
Indo-Pak war of 1965, 6
Indus water resources, 3
Internal situation, 5
Internationalising, 77

P. Pandey, *India Bangladesh Domestic Politics*,
DOI 10.1007/978-981-10-2371-2

J
Jamaat-e-Islami, 10
Jatiya Rakkhi Bahini, 10
Jatiyo Sangshad, 101
Jogighopa, 12
Joint rivers commission, 8

K
Kashmir conflict, 3
Khan, Ayub, 4
Kosygin, Alexei, 4

L
Low water level, 3

M
Mahananda, 27
Man Singh Committee, 41
Maulana Bhashani, 9
Meghna, 2
Military regimes, 14
Ministerial level, 5
Moshtaque Ahmed, Khondakar, 7
Mujibur Rahman, 7
Muhammad Ershad, Hussain, 13
Murshidabad, 40
Muslim league, 10

N
Negotiations, 1
Nehru, Jawaharlal, 4
Nepal, 2
New Delhi, 1
Non-Congress government, 11

P
Padma, 2
Pakistan, 1
Partition, 4
Partition of Bengal, 39
Politicization, 1
Port, 41
Post-1971, 1

R
Rahman, Reaz, 12
Rahman, Sheikh Mujibur, 7
Ram, Jagjivan, 11
Rahman, Zia-ur, 9
Riparian countries, 2

S
Secretary level meetings, 5
Serniabat, Abdur Rab, 8
Shastri, Lal Bahadur, 4
Siddiqui, Kader, 10
South Asia, 4
South Asian Association for Regional Cooperation, 69
South-east direction, 2
Soviet union, 4
Subansiri, 12
Sundari mangrove, 28
Swaran Singh, 6

T
Tashkent agreement, 4
Technical experts, 3
Teesta interim agreement, 123
Ten-day period, 12
Ten-day schedules, 12
Tipaimukh, 23
Track-II diplomacy, 98
Trinamool Congress, 122

U
Upstream, 2

Z
Zia, Khaleda, 18

MIX
Papier aus verantwortungsvollen Quellen
Paper from responsible sources
FSC® C105338

If you have any concerns about our products,
you can contact us on
ProductSafety@springernature.com

In case Publisher is established outside the EU,
the EU authorized representative is:
Springer Nature Customer Service Center GmbH
Europaplatz 3, 69115 Heidelberg, Germany

Printed by Libri Plureos GmbH
in Hamburg, Germany